Packaging

Packaging

A Series of Reference Books

1. Packaging: Specifications, Purchasing, and Quality Control, Third Edition, Revised and Expanded, *by Edmund A. Leonard*

Other Volumes in Preparation

Packaging

Specifications, Purchasing, and Quality Control
Third Edition, Revised and Expanded

Edmund A. Leonard
Yonkers, New York

MARCEL DEKKER, INC.　　　　　　　　　New York and Basel

Library of Congress Cataloging-in-Publication Data

Leonard, Edmund A.
  Packaging : specifications, purchasing, and quality control.

  Bibliography, p.
  Includes index.
  1. Packaging. I. Title.
TS195.L46 1987   688.8   86-23974
ISBN 0-8247-7729-8

The first and second editions were published by Modern Packaging—Morgan Grampian, Inc.

Copyright © 1987 by Marcel Dekker, Inc. All rights reserved

Neither this book nor any part may be reproduced or transmitted in any form or by any means, electronic or mechanical, including photocopying, microfilming, and recording, or by any information storage and retrieval system, without permission in writing from the publisher.

MARCEL DEKKER, INC.
270 Madison Avenue, New York, New York 10016

Current printing (last digit):
10 9 8 7 6 5 4 3 2 1

PRINTED IN THE UNITED STATES OF AMERICA

Foreword

Packaging is an immense, growing and often exciting field, but it is not all glamour and achievement. On a day-to-day basis its success depends on wise and effective execution of professional procedures, including a great deal of attention to detail.

There has thus always been a need for how-to books in packaging to present the learning experiences and problem-solving solutions that professional packagers have wrested from the field as the result of great effort, much cost and the development of considerable packaging skill.

The work that follows is such a book. It is, in effect, a road map showing how to chart a course through the complex steps of determining what packages you want, communicating these needs to suppliers so that they will understand them, and maintaining control over the quality and uniformity of supplies and services received. There is a very big dollars-and-cents aspect to the packaging know-how provided in the book, but more than that is the provision made for repeating package success on an orderly basis and for enhancing your own skills and knowledge as a professional packager.

The author has written this book on the basis of long and practical experience. He was Packaging Development Manager of General Foods Corp. He is now a packaging consultant and Adjunct Professor at Cornell University.

FOREWORD

Mr. Leonard was in 1972 president of The Packaging Institute and has served the organization in many capacities, including those of director, treasurer, and vice president for educational activities. His other professional activities include membership in the World Packaging Organization and vice presidency in charge of the American Management Association's Packaging Division. He is a director of the Packaging Education Foundation. He is an honorary member of the Australian Institute of Packaging and appeared as a guest speaker at the Conference of the British Institute of Packaging in 1966, the National Packaging Association of Australia in 1967, World Packaging Congresses in Tokyo 1972, Sao Paulo 1974, Manila 1975, Chicago 1978, and Paris 1984.

He has two degrees from Columbia University, B.A. in Chemical Engineering, M.A. in Industrial Chemistry, and has done graduate work at the University of Colorado.

His published works include more than 60 articles and papers of a technical and marketing nature on textiles, packaging, water pollution control, and economics.

William C. Simms
(Deceased December 1984)
Modern Packaging Encyclopedia and
Modern Packaging Guidebook Series

Preface

The development of new and improved packages is a busy, varied, creative, and intensely interesting occupation. As a discipline, it demands forethought in planning, meticulous care in execution, and perseverance in completion of details. A sprinkling of innovation is the spice in every development which adds the reward of doing something that has not been done before.

It must be recognized, however, that development is not an objective in itself. As in a game, the play can be fascinating, but the object is to win. A packaging development must result in something useful that contributes to the protection, distribution, and marketing of a product. To make such a contribution requires that the result of each development be communicated to those who can use it, with assurance that it will perform as intended.

Specifications and quality assurance programs are the classic tools for linking developments to production. The link is sometimes not as strong as it should be, possibly because specs and quality control programs are "dry" subjects, tedious to write and difficult to read. The developer has little interest in detailing on paper all the facts that may be important, but which are no longer new to him; those who are to use the development cannot do the writing because they have neither the facts nor the knowledge of which facts are important.

This book has the purpose of making the transition steps between development and production a little more acceptable, understandable, and interesting. Specifications and purchasing and quality control are those steps; in the chapters that follow, we shall treat these three topics as media for communicating information and basing decisions.

The parts of this book that treat purchasing do so in the context of that function as part of the communication network needed to describe, obtain, and use with confidence the packaging materials most commonly applied commercially today. This is not a manual on Purchasing which, beyond the technical communication activity on which we focus here, must apply skills and techniques of bidding, negotiating, pricing, logistics, credit, and finance. All these are capably covered in other sources.

The emphasis in this book on specifications and quality control is aimed toward simple and understandable principles, with specific examples to show how they may be applied. The technical depth on quality control is intended to be sufficient for manufacturing application by nonmathematics specialists. The bibliography includes references to more technical works.

The first edition, published in 1971, was written as the text for an eight-lecture course of The Packaging Institute and used across the country by its chapters in their continuing education programs. The second edition, published in 1976, updated the first and included a new chapter on Unitized Constructions for Distribution, recognizing the rapidly growing diversity of shipping modules and the options for stabilizing them. The Society of Packaging and Handling Engineers added the title to their list of references for study in preparation for its Certified Professional examination. The second edition was also chosen as the textbook for several college courses in packaging technology.

In this third edition, everything is again updated, a tenth chapter on Computerized Specifications is added, and coverage of coextrusion is inserted appropriately. Modern constructions such as metallized films, paper, and board are added in Chapter 8.

Opinions expressed or implied in the text are those of the author alone and do not claim to represent the position of any organization.

I would like to thank Mr. Harry Susla of Latchford Glass Company for his guidance on glass specifications; Messrs. Ron Dembowski of Ball Plastics and Ed Martin of Printpack for demonstrations of coextrusion structural analytic methods; and Professor Jim Goff of Michigan State for his pioneering papers on identification of films by infrared transmission.

Edmund A. Leonard

Contents

Foreword (William C. Simms)		*iii*
Preface		*v*
1.	Functional and Marketing Criteria in Specifications	1
2.	Manufacturing and Cost Criteria in Specifications	29
3.	The Buyer-Supplier Relationship	45
4.	Quality-Control Principles	57
5.	Glass Package Specifications and Quality Control	81
6.	Metal Container Specifications and Quality Control	111
7.	Plastics Package Specifications and Quality Control	131
8.	Paper, Board, and Flexible Package Specifications and Quality Control	153
9.	Unitized Constructions for Distribution	183
10.	Computerized Specifications	203
	Appendix: Technical and Trade Associations	219
	Bibliography	221
Index		*227*

1

Functional and Marketing Criteria in Specifications

> ... A spec is not an epitaph written at the end of development, but a working communication for the use of the thing developed....

I. INTRODUCTION

The fact of life underlying our title and subject is that new packages are continually being developed, moving from the idea stage through several necessary steps into production; and developed packages are continually being refined in appearance, cost, and function. Major or minor changes are being studied or put into effect at any given time on most packaging. It is not unusual for all the packages of a large product manufacturer to be out of date in 1 year's span.

Such a stream of innovations and changes requires a system of controls to guide it on an orderly course. That system is provided by a management attitude that relates procedures for developing packaging specifications, purchasing the specified materials, and controlling their quality. Viewed sequentially, the first steps in development of a new or changed package end with the writing of a specification; the latter is then used by a purchasing agent to negotiate supplies of the specified material, and finally, a third group of persons takes responsibility for assuring quality, or conformance to the specification.

The theme that will be developed in the pages to follow is that a specification should be written not as a monument to the genius of its creator, but as a working communication that can be used by a buyer with his suppliers, and which allows manufacturing and quality control people to do their jobs, by including quality criteria and levels required in the delivered materials.

Stated thus in a few words, the whole process sounds simple enough, but the real problem is the most difficult in all human affairs: that is, to communicate true understanding, in this case among dozens of people involved at the decision-making level in purchasing, selling, manufacturing, delivering, examining, and using a specified packaging material. It is a safe bet that everyone who reads this sentence will be able to recall a fiasco that occurred because someone "didn't get the word."

A good step in communicating for understanding is to use a common language, and a few definitions will be helpful:

1. *Specification*: a *description*, usually with dimensions and illustrations, of an article or process.
2. *Packaging Material Specification*: the description of a package, or a package component, or the raw material from which it is made.
3. *Package Component*: a *part* of a complete package, such as a can end or a paper label.
4. *Packing Specification*: the description of a *process or procedure* for filling and sealing a given product into its package.
5. *Criterion*: a *requirement*. (Plural: criteria.) For example, a criterion for most retail packages is that they have a light-colored spot to permit legible price marking.
6. *Package Structure*: the *materials* of which a package is made, and their dimensions.
7. *Package Function*: the *services* that a package performs,

FUNCTIONAL AND MARKETING CRITERIA 3

starting with containment and isolation of a desired amount of product, usually including protection and identification of that product, and possibly such additional criteria as easy opening, reclosure, and disposability.
8. *Package Graphics*: the *surface decoration* of a package with copy and illustration or pattern, including considerations of color, reflectance, and gloss.
9. *Purchasing*: all of the activities concerned with *procurement* (of packaging supplies), including the selection and evaluation of suppliers, obtaining and evaluating bids and proposals, negotiating prices and other conditions of buying and payment, contracting and ordering, forecasting supply problems and opportunities.
10. *Quality Control*: all of the activities concerned with keeping products, packaging materials, and packing processes in *compliance* with their respective specifications, so that they are acceptable according to the criteria set for them.

To make life simpler and in closer agreement with on-the-job language, a few abbreviations for some of the long words are desirable:

1. *spec* for specification,
2. *q.c.* for quality control,
3. *p.a.* for purchasing agent; *buyer* is short but not quite synonymous, being frequently the title for a job that reports to the purchasing agent.

The development of our subject is now best undertaken with a detailed study of the spec. *A good packaging material spec is a description accurate and detailed enough that a p.a. can ask for bids on it, and that a supplier can quote, manufacture, and deliver material having the appearance and performance intended by the writer of the spec.* The writer's task, of course, is to decide what to include, how to make it understandable, and how to keep it from becoming unbearably long and forbidding. To specify some complex packages, a combination of completeness, clarity, and brevity is an unattainable ideal; a happy balance is as much an art as a science. If completeness is necessary, brevity must suffer and clarity is endangered, while if brevity is most important, clarity is not hard to achieve, but completeness is risked.

A handle can be put onto this circular dilemma by reviewing the usual course of events in development of a spec:

Step 1. A packaging engineer obtains and tests laboratory samples, refining and retesting them until he is satisfied with the results, or until other pressures force him to a conclusion.

Step 2. He writes a spec draft, including everything he considers important.

Step 3. The draft is reviewed in-company with those who will buy the material in question, those who will handle it in production, and those who will be responsible for q.c. The draft is modified and amplified to a point of agreement.

Step 4. The p.a. submits the draft to one or more vendors for bids and comment. The latter respond with questions, proposals, and notice of any conflicts with their own practices and capabilities. After due process and discussion, sample quantities from one or two tentative suppliers are tested for acceptability.

Step 5. A third draft of the spec is made based on the appearance, structure, and performance of tested and approved samples. This draft probably includes refinements of detail over the first two and is established as a tentative standard for first production.

Step 6. After the first production experience, the "formal" spec is written, including all elements found necessary and desirable to date. Unexpected variables discovered to be important in later production may require additional refinements in the spec.

What has actually happened is that the packer wrote a spec describing what he thought he wanted; the supplier delivered what he thought the client could use, based on what he could make; and after some trial and error the two found a common ground of agreement. The formal spec then became an historical record of what was tried and worked, of possible value for starting up additional suppliers when and if needed, and for the information

FUNCTIONAL AND MARKETING CRITERIA

of personnel who later take the burdens of those on the original development team.

Even for the latter two uses, the formal spec usually becomes a platform for further alteration. A second or third supplier, for example, will have manufacturing equipment and practices somewhat different from the first, which will make it impossible or undesirable for him to deliver the identical spec, but so long as it will operate interchangeably with the original, it can be an acceptable alternate or equivalent spec.

As for further developments, it is inevitable that new technology and/or marketing evaluations will lead to improvements of the formal spec in cost or function or appearance, perhaps by employing a new material of construction, or lightweighting, or strengthening for higher-speed packing, or new graphics.

Looking back over this "usual" process, it can now be asked whether this is the proper way to develop a packaging spec, and how does it solve the dilemma of balancing brevity, clarity, and completeness? A bit of reflection leads to the following observations and conclusions:

1. The only substitute for developing a spec through the six steps described is experience — so much experience, in fact, that the writer knows all the answers at the end of step 1. Note that he must be able to speak for both his and the supplier's capability, assuming that the latter has been identified at that early stage.
Since such experience is obviously rare, there is no choice but to go through the several steps in spec development.
2. A spec is continually evolving. Even the formal spec is subject to technical, cost, and marketing improvements as long as it remains commercial.
3. This being the case, there is no such thing as a "complete" spec. If it existed, it would be too long and detailed, it would have cost too much to prepare, it would soon be out of date, and it would cost too much to maintain up to date.

This philosophical examination leads us to refine our underlined definition of a good packaging spec by adding the condition that it be *"sufficiently detailed to minimize the risk of failure to perform, and sufficiently flexible to permit change and improvement."* This refinement solves our brevity-clarity-completeness dilemma only by recognizing that the spec cannot be the sole communication among its writer, p.a., supplier, and q.c. people. The other forms of communication needed for full understanding are treated in later chapters.

There is a temptation to call the formal spec "final" and to hope that the drafts which precede it would have proven to be "final," but it never comes to pass, even in the simplest basic materials; and in the author's opinion, it is not desirable to seek finality, since it tends to form a mental block against future improvement. No spec can last indefinitely, and this should be no surprise, because the situation is no different with other forms of "permanent" documents. Laws, constitutions, contracts, treaties, scientific theories, all are continually evolving in response to economic and social human needs. The cause of progress is best served by using specs, like the others, as foundations for further improvement and not as restrictions against change.

Now that it has been considered how long or detailed a spec should be, the next consideration can be what it should include qualitatively. Regardless how long or short, a good packaging material spec should leave no important aspect of the package to the imagination: structure, product protection, cost, packing performance, consumer function, and graphics. These can conveniently be grouped into two sets of criteria.

1. The "outside world's view" of the package, which is composed of the marketing criteria — product protection, consumer function, and graphic treatment.
2. The product manufacturer's view of the package, which includes the factors of cost, packing-line performance, and product identification.

This chapter concentrates on the "outside view" of specifications, describing those entries which assure coverage of marketing. objectives. The next chapter will deal with the cost and manufacturing side of a spec.

Figure 1 is a brief and simple outline of a specification form. Items I and II relate to manufacturing operations in the packer's

SPECIFICATION FOR _____ NO. _____

PROPERTY OF XYZ COMPANY   EFFECTIVE DATE _____

I.   SCOPE: This specification describes _____ package for _____ product. It is for use at _____ plant(s), and supersedes No. _____ dated _____.

II. CONSTRUCTION:
   A. Materials: Weight, thickness, stiffness, grade, etc., with limits.
      Size — prints to be attached, with dimensions and limits.
      Special features — such as pour spout, tear tape or string, notches, opening devices, etc.
   B. Assembly: List of steps in packing, with charts if required.

III. PRODUCT PROTECTION:
   A. Climatic: Product shall not (cake, rust, sift, freeze, dry out, etc.) when packed and stored for _____ period of time under _____ conditions of temperature and humidity. Test methods appended.
   B. Mechanical: Product shall not (dent, scratch, crack, crumble, leak, etc.) when impacted in package under _____ conditions. Test methods.

IV. CONSUMER FUNCTION: The package shall open readily without need of tools when the (opening device) is (pulled, turned, torn, etc.) with a force not to exceed _____ (ounces, pounds, in-lb., etc.) of force, and shall reclose (if appropriate) under _____ conditions. The label shall not wash or rub off under normal use as determined by _____ test. After opening and reclosure, although normal storage position is upright, the package shall not (leak or sift) when stored on its side, etc., etc.

V.   GRAPHICS: Mechanical art and color numbers or swatches specify the decorative treatment. Scuff and wash resistance shall be _____. Test methods appended or referred.

Figure 1   Specification outline.

plant, while III, IV, and V speak to the marketing objectives of the package. It is true that the structure of the package, stated in II, should guarantee its adequacy for III and IV, but III clearly states the *intent* of the package spec. Its value is this: A legalistic description of structure cannot completely convey the purpose of that structure, while a simple statement which says what the package must do notifies the supplier very clearly of his responsibility.

During the several steps of developing the spec, a supplier can make proposals for the structure, given the instructions in III and IV, but without that input he can only blindly follow the structural spec and hope that it does what the buyer wants of it. There is a most important principle here implied: no one knows the packer's business as well as the packer; he is, therefore, the one best qualified to say what the package must do. And on the other hand, no one knows how to make the package as well as the supplier, and the latter can therefore best define the structure. The two parties must communicate their respective expertise until they reach agreement that the structure proposed by the supplier matches the performance expected by the packer. It may be asked why the structure should be in the packer's specification at all, if he can only contribute functional requirements. The answer is that structure is easier and faster to measure nondestructively in auditing deliveries of the packing material. For example, the breakage resistance of a carload of glass bottles cannot be checked without breaking a sizable sample, but wall thickness and weight can be checked without breaking any, and those factors are the principal structural determinants of strength. Stated another way, which will be treated in more detail in a later chapter on quality control, *breakage resistance* is an *attribute* or *property* difficult to measure objectively, while *thickness* and *weight* are *variables* or *quantities* easy to measure numerically. This is an example of planning the spec to coordinate with those who must use it, in this case the q.c. people. Another example with the same purpose is the indicated inclusion of test methods, which should be established by the time the spec becomes effective.

The format of a specification cannot avoid great dependency on the nature of the article or material specified. The tools for

FUNCTIONAL AND MARKETING CRITERIA

building specs are words, pictures, and numbers; the skill and experience of the writer must be employed to combine them so as to omit nothing important, and to include nothing unnecessary. There are some entries which all specs should have, however, and these are referring to Figure 1:

SPECIFICATION FOR____

The material, component, or package described must be named at the beginning or it requires an annoying search for the subject of the spec. It may be "Folding Carton," or "Corrugated Box" or "48 mm. Metal Cap" or "Carton-Sealing Glue," for example, but it identifies the class of material at the outset.

NO.____

A numbering system for specs is indispensable. It allows indexing, coding for data processing in cost and inventory controls, notation for reference on product specs or process specs, and for simply clarifying the fact that one folding-carton spec, for instance, is distinct from another. A deceivingly simple question arises — *How* do you number specs? All we need do is consider the alternatives, and more questions pop up:

Alternative	Question
1. Start from #1 and proceed consecutively, regardless of package type.	How do you find a spec in a file without a separate index and cross-reference?
2. Classify specs by package type with a code letter prefix, such as C-1 for a carton spec, S-1 for a shipper spec, B-1 for a bottle spec, and so on.	Better than above, but within package types having large numbers of specs, how do you find one in question?

Alternative	Question
3. Use a combination of several code designations ahead of the spec number to narrow the field of search, such as G12J-1 for a Glass that holds 12 ounces of Jelly, Spec #1.	What do you do when the same jar is also used for 16 ounces of honey, or for 8 ounces of chopped dried glace fruits — apply three spec numbers to the same jar?
4. Classify specs by brand or product name or product code.	When one package is used for two or more products, do you duplicate specs under as many products?

It will be noticed that the several alternatives fall into one of a few schemes:

1. Numbering package specs chronologically as issued, without regard either to package type or product classification
2. Classifying by package type without regard to product
3. Classifying by product without separation by package type
4. Classifying by some combination of the above possibilities

Is one of these choices best, or better? Are there other possibilities? Let us consider the pros and cons, with illustrative examples. (See Table 1.)

In this example, the hardware distributor found need for a little more than 1500 specifications over a period of 15 years. To keep the spec file or book up to date requires that superseded specs be removed or overprinted, "SUPERSEDED BY #____," or they may be inadvertently ordered by a new foreman or buyer. The major fault of this system is that it is not easy to find the package spec for a given product. One must scan the list of specs, starting with the latest, or set up a cross-index by product, and maintain the latest package spec number listed beside it. There is a chance for error if someone has a cross-index which is not up to date on specs issued — he may fail to cross off a superseded spec number and use obsolete materials. Confusion can also arise when reference is made in one spec to another that has been superseded. For instance, it is noted in Spec #409 that it is an alternate for #1, but the latter was superseded by #1039. In this

FUNCTIONAL AND MARKETING CRITERIA

Table 1 Chronological Numbering (Examples Are Hardware Packages)

Spec No.		Date Issued
1	Reverse-tuck carton for 1/2 lb. 4-penny wire nails and for 100 1/2 inch flat-head screws, and for 100 upholstery tacks, and for 1 pair 2-inch brass hinges.	Jan. 1970
2	Reverse-tuck carton for 2 lb 4-penny wire nails	Mar. 1970
3	Reverse-tuck carton for 1 lb 6-penny wire nails	Mar. 1970
409	Blister pack on card, alternate for Spec 1 when distributed for rack merchandising	Jun. 1975
410	Blister pack on card, alternate for Specs 2 and 3 when distributed for rack merchandising.	Jun. 1975
790	Same as 1, but with cellophane overwrap for new product, 1/2 lb wire brads	Sep. 1978
1039	Supersedes 1 by replacing gray board stock with lighter-weight bleached white board	May 1983
1040	Supersedes 2 for same reason	May 1983
1041	Supersedes 3 for same reason	May 1983
1042	Supersedes 790 for same reason	May 1983
1508	Plastic bag with stapled header, 1 punched hole, as alternate package to #409 for 1 pair 2-inch brass hinges *only*, a variety store package	Feb. 1985

system of spec numbering, the spec writer(s) must use discipline to avoid referrals that can cause later confusion, but such avoidance weakens the coverage of intent that is important to a spec. The one referral that must be included, of course, is notice that the new spec supersedes an old one. Figure 1 provides for this referral.

II. CLASSIFICATION BY PACKAGE TYPE (SAME EXAMPLES)

In this system a series of code symbols, letters or numbers, is set up for the several package types which are used, followed by a numbering system which is sequential for its package type. Let us assume the following package codes for the hardware packages above:

C = Carton,
B = Blister pack,
OC = Overwrapped carton,
PB = Plastic bag

Then we can arrange, in Table 2, a comparison of this system with that above which uses simple chronological numbering.

Table 2 Comparison, Two-Spec Identification Systems

Chronological system			Package-type classification	
Spec no.	Supersedes	Date Issued	Spec no.	Supersedes
1	—	Jan 1970	C-1	—
2	—	Mar 1970	C-2	—
3	—	Mar 1970	C-3	—
409	—	Jun 1975	B-1	—
410	—	Jun 1975	B-2	—
790	—	Sep 1978	OC-1	—
1039	1	May 1983	C-565	C-1
1040	2	May 1983	C-566	C-2
1041	3	May 1983	C-567	C-3
1042	790	May 1983	OC-29	OC-1
1508	—	Feb 1985	PB-1	—

FUNCTIONAL AND MARKETING CRITERIA

A comparison of the two systems permits the following observations:

1. The chronological system index tells how many active specs there are at any given time, but gives no idea what kinds of packages they are.
2. The package-type system tells how many active specs there are in total and by package type.
3. Both systems require maintenance of an index, since the numbers in both have no significance other than the order in which issued.

III. PRODUCT CLASSIFICATION (SAME EXAMPLES)

Here the products must be identified by codes, and the package spec becomes part of the product spec. For example, the limited amount of information in the hardware products we are studying can be coded in four-digit numbers. Let us postulate the following codes:

Fourth decile product type	Third decile product size	Second decile pack size	First decile package type
4=nails	4=4-penny	5=1/2 pound	2=carton, nat'l
1=screws, flat	5=1/2 inch	0=100	3=blister
6=tacks	7=upholstery	1=1 pound	4=o'wrap carton
8=hinges	2=2-inch	2=a pair	5=carton, bleached
	6=6-penny		6=plastic bag
	8=brads		

With these rules established, it is possible to arrange a spec table for the same hardware products, comparing them to the previous two systems. (See Table 3.)

Comments can now be made on the product-spec system versus the other two.

1. More specs in total are required because duplication cannot be eliminated where one package is used for more than one product.

Table 3 Comparison, Three-Spec Identification Systems

Chronological system	Package-type system	Product-spec system	
Spec no.	Spec no.	Spec no.	Supersedes
1	C-1	4452	—
1	C-1	1502	—
1	C-1	6702	—
1	C-1	8222	—
2	C-2	4412	—
3	C-3	4612	—
409	B-1	4453	—
409	B-1	1503	—
409	B-1	6703	—
409	B-1	8223	—
410	B-2	4413	—
410	B-2	4613	—
790	OC-1	4854	—
1039	C-565	4455	4452
1039	C-565	1505	1502
1039	C-565	6705	6702
1039	C-565	8225	8222
1040	C-566	4415	4412
1041	C-567	4615	4612
1042	OC-29	4855	4854
1508	PB-1	8226	—

2. A one-digit code is not very specific for designation of package type. Three different carton specs all have the same notation as 2, two different blister packs as 3, and three different bleached cartons as 5. Even the total of 10 single digits does not allow for many package types. For any reasonable line of products at least three digits would be needed for package designations alone to allow for a maximum of 1000 package specs.

FUNCATIONAL AND MARKETING CRITERIA 15

3. The total spec has to be replaced if either the product or the package designations change. Both this and the first comment above mean that with this system the index and cross-references would be longer and changes would be more frequent.
4. This code system is easier to "read" than either of the other two, since each digit describes a feature of the product or package, while the digits in the other two systems have no such meaning, merely separating one spec from others. The product-spec system is more likely to prevent errors, therefore, when everyone in the operation knows, for instance, that *4455* means *nails, 4-penny, 1/2 pound, in a bleached-board carton.* To learn that much, one would have to look at Spec No. 1039 in the chronological system or C-565 in the package-type system.
5. The product-spec system is well suited for data-processing applications to the total company business. For example, the code *4400* would apply to all production and bulk inventory data on 4-penny nails, while sales and packaged inventories can be totaled from the data on codes *4453*, *4413*, *4455*, and *4415*. In comparison to this advantage alone, the shortcomings noted in the first three comments become insignificant.

IV. MIXED CLASSIFICATION (SAME EXAMPLES)

It is easy to see that either the chronological or the package-type system can be combined with the product-spec system by adding three more digits. This makes a code whose first three digits describe the product, and last four the package. For example, the code *4854* (1/2 pound brads in a bleached-board carton, cellophane overwrapped) when combined with the chronological system would be 485/1042 or 4851042. The second 4 is eliminated from the combined code because it becomes redundant. If 4854 is combined with the package-type system, the code becomes 485/5029 or 4855029. In this case the second 5 is retained because it replaces the abbreviation *OC*, and *29* becomes 029 to provide for a thousand specs within the class of overwrapped cartons. Of the two combinations, the latter is preferable, since more information can be "read" from it. The last three digits can also be converted, of course, from a mere sequential function to a coding pattern, to make the system still more meaningful.

The use of word- and data-processing systems to aid the control and coordination of business operations is heavy and increasing. These systems can and do treat purchasing, production, costs, inventories, distribution, and sales data in one uniform method, and if specifications can be identified by compatible nomenclature, they can be better integrated with the rest of the company operations.

Since the several data-processing systems in use vary considerably, there is no point in probing the subject more deeply here of how to number specifications. Each system user needs to develop his own detailed procedures, and we can best summarize this discussion with the following suggestions:

1. A strictly chronological numbering system is meaningless without the detailed specs and isolated from company systems. Better systems are available, and this one has little to recommend it other than that it can grow indefinitely and requires no planning.
2. Combinations of letters and numbers are compatible with most word-processing systems, and therefore can be recommended for spec identification, when the 10 digits will not suffice for the variables in a code position.
3. *Coded* alpha-numeric systems (where the positions have meaning rather than mere sequence) are the most flexible spec identification systems and the easiest to "read." They are especially useful if the company uses data processing, since they can be used to make specs compatible with the rest of the company's system. More will be said on this matter in Chapter 10, "Computerized Specifications."

Referring back to Figure 1, we examine further some elements of a specification.

DATE____

FUNCTIONAL AND MARKETING CRITERIA

Most specs are *issued* or *become effective* as of a certain date, and this date should appear on the spec. The reasons for dating a spec are several:

1. During the package development, several tentative specs may be issued, and they may all carry the same number. The latest date, of course, would be the spec still active.
2. Costs may get out of line if a given spec is kept unchanged for a long time, say 2 years, without being reviewed. A routine survey of dates allows the selection of the oldest as the most likely candidates for cost reduction, or for improvement of function.
3. The same can be said about market position. A spec that is more than 2 years old may be surpassed by competition in consumer appeal. Dated specs can be scheduled for comparison against competition at any desired interval.
4. Specs are often given to suppliers who have no knowledge of the code in the buyer's spec numbering system. The effective date or date issued, in combination with the title, has the most meaning to the supplier for his own internal communications.
5. Files of old specs lose their value unless dated. Over a period of 10 years an equal number of specs may have been used for a given product. Unless they are dated, it is soon forgotten which was used when. Use for this information arises with surprising frequency, such as for evidence in patent litigation, for patent application by establishing first date of commercial use, for negotiating with a supplier on liquidation of obsolete packaging materials inventories, or simply for reference in writing company history.

PROPERTY OF XYZ COMPANY

If specs were used only for internal purposes, there would be no need for company identification, but this is almost never the case. A spec is a valuable piece of property, which probably cost thousands of dollars to develop, and in the case of consumer products a spec probably has competitive value, too. It should be protected at least by stating whose property it is and, when given to a supplier for his use in supplying the packaging material, it may wisely be stamped CONFIDENTIAL or NOT TO BE COPIED or NOT TO BE RELEASED TO UNAUTHORIZED PERSONS.

I. SCOPE

The introductory sentence states briefly what the spec covers, and adds to the title, spec number, and effective date a few words on what product the package is for, at what packing location(s) it is used, and what earlier spec, if any, it supersedes. This information reinforces the spec number, if it is coded, or communicates it to a supplier who does not know the code, and advises him what plant he is to supply. The inclusion of a short scope statement is very useful to orient the reader on what the spec is all about.

III. PRODUCT PROTECTION

The main subject of this chapter is the functional and marketing criteria for a package, and how to write them into a spec. With the preliminaries on general format disposed of, we may proceed to consider the first function of a package, which is product protection.

Almost all goods, agricultural and manufactured, pass through similar cycles in their distribution: immediately after packing, they are warehoused; then transported to a distribution center or customer warehouse; then moved to a retail outlet; removed from the shippers; displayed as individual packages; sold and carried to their final destination by the purchaser.

During this cycle, the products need two kinds of protection — climatic and mechanical. Climatic protection is that property of

FUNCTIONAL AND MARKETING CRITERIA   19

the package which resists damage to the product by atmospheric effects, such as light, heat or cold, dampness or dryness, and oxidation, for example. Mechanical protection is resistance to physical damage, such as scuffing, chipping, breakage, scratching, or puncturing.

The needs for product protection of both kinds must be determined at the time when the product is developed. This is the business of the product manufacturer and packer; the packaging materials supplier can be responsible only in an assisting capacity. In many cases, to be sure, a supplier can run all necessary tests and recommend a package, especially if it is to contain a product with which he has had experience. For instance, if the XYZ Company develops a line of preserves, a glass container manufacturer can readily suggest packages which will provide all necessary protection. He could pack samples and run abuse tests for breakage resistance and on these results hand a spec to the packer. Strictly speaking, however, the packer should make his own taste tests after storage and should also test damage resistance through his own distribution, since there may be something unique about his product formulation and distribution pattern that will not be picked up in suppliers' tests.

It is a general fact that no one knows his product so well as the manufacturer, and no one else is so well qualified to evaluate its performance. The intent of the package's protective function may be stated here in as much detail as the writer of the spec considers necessary; the extent of such detail will probably depend on the value of the product contained. The term "protection against the elements" is often used; let us identify those elements, which will provide a checklist for the use of the spec writer. (See Table 4.)

Climatic influences and mechanical forces occur independently and can therefore be considered separately, but contamination is usually the result of both. The bacteria, molds, and insects that contaminate are part of the climate or environment, but they are allowed to enter a package when mechanical forces cause a break in the barrier. They can also, of course, enter if the package was not sealed tightly when assembled. Pilferage is separate and distinct — a willful human act of tampering, but nevertheless a factor which can be foreseen and provided for in a package

Table 4 Elements Resisted by Protective Packaging

A. Climatic influences	B. Mechanical forces
Temperature	Impact
Light	Abrasion
Humidity	Acceleration
Gas reactions	Vibration
Package reactions	Torsion
C. Contamination	
Bacteria	
Molds	
Insects	
D. Pilferage	

design. Examples of statements in a spec which relate to each of the above can be cited:

1. Temperature resistance
   a. "The shipping container for frozen whole blood must carry 10 pints, and when packed with 20 pounds of dry ice, shall not allow the internal temperature to rise above -40°F in 72 hours, when the unit is stored in a room at 105°F."
   b. "Carbonated beverage bottles shall not burst when filled to normal levels with 3-volume product and stored at temperatures up to 150°F for 24 hours."
   c. "Plastic dispenser packages for PVA emulsion glue are distributed by truck and warehoused nationally. They are stacked in trucks two pallets high and warehoused 15 feet high, in their corrugated shippers with partitions. It is intended that in transportation and storage up to 120°F exposure, the individual dispensers shall not soften and leak or distort under static loads up to 2 pounds each."
2. Light resistance
   a. "Photographic paper in envelope packs of 25 sheets

shall be light-tight in the normal distribution cycle for a shelf-life of 1-1/2 years. This may be evaluated by exposing envelope packs in their shipper to a standard laboratory rough-handling procedure; then removing 1 pack from each of 10 cases and exposing them to 1/2 hour in a Fadeometer, half with the front face and half with the flap side of the envelope facing the arc light. The top and bottom sheets in each pack shall be processed in the darkroom, and if they show any sign of light strike, the envelopes shall be declared defective."
   b. "The milk bottles shall absorb 98% of incident ultraviolet radiation in direct sunlight, such that the vitamin assay of milk exposed in the bottles 4 hours between 9 A.M. and 4 P.M. of a clear sunny day shall be not less than 90% of its original assay."
3. Humidity
   a. "Frozen orange-juice concentrate in retail 12-ounce packages is required to have a shelf life not less than one year when stored at $0°F$ in a saturated atmosphere. It is necessary that this (e.g., spiral-wound easy-open) package limit product weight loss to less than 2% under the above condition."
   b. "XYZ Dry Patching Plaster is hygroscopic and will cake if its moisture content exceeds 6% by weight. It is packed at a maximum of 3%, and its package must prevent pickup of more than 3% over a period of one year in any part of the continental United States."
4. Gas reactions
   a. "Oil-based varnish will gel and harden if air (oxygen) leaks into the can. The double-friction lid on pint and quart cans and the screw closure on gallon cans must be airtight, such that no viscosity increase or skin for-mation takes place within 6 months storage at room temperature."
   b. "Carbon dioxide in air will react with certain dyes on acetate fabrics. The plastic bag packages for XYZ acetate blouses shall have airtight seals and shall resist permeation of carbon dioxide such that no color change will take place in packaged blouses stored 3 months in an atmosphere containing 1% carbon dioxide."

5. Package interactions
   a. "Cans for XYZ latex paint shall be coated internally with enamel linings to prevent rusting for a period not less than 1 year of room-temperature storage filled with any color of paint."
   b. "The plastic linings of paperboard milk cartons shall be completely odor free and shall not add off-flavors to milk packed and distributed in these cartons."
   c. "Bags for potato chips must be formed of grease-resistant paper stock, such that no oil staining is visible on the outer surface of the bags after storage for 2 weeks filled with product. Also, the adhesion of printing ink on the outside of the bags shall not be reduced by grease picked up on the inside surface, as measured by a scuff-resistance test."
6. Impact resistance: "One-quart fishbowls in cases of 36 shall not break when shaken for 1 hour at 1 g on an L.A.B. vibrator and impacted on an inclined-impact tester for a cumulative total of 150-foot falls."
7. Abrasion: "The paper labels on XYZ brands of canned fruits and vegetables shall not show scuffing during any normal conditions of distribution and storage. This may be evaluated by examination of packed goods in case lots before and after a 1000-mile truck shipping test, and also by standard laboratory scuff test on new labels aged at least 24 hours after application of the overprint lacquer or varnish coat."
8. Acceleration: "Shipping cases for XYZ canned fruits and vegetables shall be coated on top and bottom surfaces with silica or other frictionizing material or device, such that two cases filled with product and stacked will not slide when tilted to an angle of 10° or accelerated at 1/8 g (4.0 ft/sec/sec)."
9. Vibration: "Glass jars packed with XYZ powdered chocolate drink mix are closed with metal screw caps applied with a torque of 25 to 30-inch pounds. During all normal conditions of distribution the caps shall not back off or loosen to the point of releasing the seals. After L.A.B. vibration for 1 hour at 1 g no caps in a case load shall show an opening torque lower than 5 inch-pounds."

FUNCTIONAL AND MARKETING CRITERIA   23

10. Torsion: "The packages for 6-foot fluorescent lighting tubes must resist deformation that could stress the glass tubes. A package for 12 tubes shall be rigid enough that when supported in the center and loaded on opposite top corners with two 10-pound weights, the opposite end panels shall be not more than 5° displaced from parallelism of their edges."
11. Contamination with bacteria: "Printed glassine paper for wrapping chocolate nut bars shall be free of any attached foreign matter and all pathogenic bacteria and spores. When cultured, the paper shall have a bacterial count lower than 5000 per square inch."
12. Contamination with molds: "ABC film used for pouch packaging of XYZ natural cheese slices shall contain sufficient antimycotic food-compatible material to prevent mould growth for a period not less than 4 weeks of storage at 45°F."
13. Insect contamination: "Glues used in forming and sealing XYZ dry food product cartons shall be entirely synthetic and unattractive to insects as food. All cartons shall have sealed waxed liners free from gaps and openings that would permit insect entry."
14. Pilferage: "A tamperproof feature is required on all XYZ instant tea jars. The paper membrane shall be glued across the mouth of each jar with an uninterrupted seal around the circumference. The membrane shall not be peelable, but shall require cutting or tearing for removal."

The 14 statements above constitute examples which together represent coverage of all the important factors in III. PRODUCT PROTECTION. In short, a maximum of about 14 statements, with necessary test methods, should suffice to describe the protective intent of any ordinary package. If the product requires attention to all of them, it can add a few pages to the spec; there are, fortunately, few volume products sensitive to most of the climatic and mechanical effects but they are important ones; in the field of perishable foods: meats, baked goods, and several dairy products. Later chapters will go into more detail on specific kinds of package specs.

IV. CONSUMER FUNCTION

We have examined what a package does for the product it contains; we now turn our attention to considering what it does for the consumer, and how to state such functions in a spec. It can be said of most consumer packages that they must provide easy access to the product in respect to opening and dispensing, easy reclosure if more than a single-use package, with product protection between successive openings until empty, and finally, easy disposal or the opportunity for reuse. A few other considerations will often apply to the packages for specific product classes; for example, food packages should be of a size and shape convenient to store in kitchen cabinets or on shelves, medicine packages should fit into bathroom cabinets, paint containers should stack in tool closets or under workbenches, cosmetic packages should stand solidly on dressers without danger of spilling and marring the furniture, purse packages should be small and not spill in any position or open unintentionally.

Since there will generally be fewer criteria for consumer function than for protective function, statements for the former may not need to be as lengthy. Fourteen protective criteria are listed above, while five criteria for consumer function (opening, dispensing, storing, reclosure, and disposal/reuse) will cover most situations. In Figure 1 the paragraph on CONSUMER FUNCTION covers all of these except disposal in a small space. It is possible that a maximum of four test methods will need to be appended for opening force, reclosure protection, label permanence, and leakage resistance in nonupright position. At this point it will be useful to illustrate with a few examples how consumer function may be written into a package spec.

Sample Specification Statements on Consumer
Function of Packages

1. Small plastic scouring-powder package
    a. "The product is used principally for removal of stains and hardwater soap deposits from sinks and bathtubs and water closets. The package is intended for multiple

FUNCTIONAL AND MARKETING CRITERIA

use, with storage convenient for kitchen and bathroom cabinet and counter top. The plastic square pressure-sensitive closure over the perforations in the cap shall be easily peeled off for initial opening. It shall lift off cleanly in one piece without tearing or shredding when pulled upward from one corner with wet or dry fingers."
  b. "The perforations in the cap shall be completely open and free of rough-cut defects that impede the free dispensing of the product."
  c. "The paper label wrapped around the middle of the package shall be easily torn off, leaving no shreds on the package, so that the container may be stored in the open and free of commercial messages."
2. Instant tea jar
  a. "The product is usually dispensed with a teaspoon into cups for hot tea, or into tumblers or a pitcher for iced tea. For initial opening, cap removal torque shall be not less than 10 inch-pounds nor more than 40. The inner-seal shall be firmly glued to the lip and require fiber tear for removal, both for product protection and for demonstration of tamper-proofness."
  b. "Market research has shown that each jar will be opened and reclosed an average of 20 times before it is emptied. The waxed pulpboard cap liner shall be an effective water-vapor barrier during reclosure at 10 inch-pounds of torque. To test the effectiveness of the reclosure, open jars with innerseals removed shall be reclosed at 10 inch-pounds after sampling and testing the contents for moisture level. The jars shall be placed in a humidity room or cabinet at 90°F. and 85% r.h. for 30 days and at the end of this time the contents tested again for moisture level. The reclosure performance shall be judged acceptable if a random sample of 30 jars, none shall have allowed the contents to increase more than 2% in moisture content."
3. Lipstick: "The plastic cover shall be removable from the case by sliding it off, and shall require a pulling force not greater than 1-1/2 pounds to effect removal. The stick may

be raised and lowered from the case by turning the base with a torque not to exceed 5 inch-pounds. When closed the cover shall not be accidentally removable, to avoid the possibility of staining other contents of a purse. This may be evaluated by tumbling a random sampling of lipsticks in a laboratory ball mill. Thirty sticks and 100 1/2-inch softwood balls shall be tumbled for 1 hour at 100 rpm in a 1-gallon mill. At the end of that time, none of the covers shall have come off."

V. GRAPHICS

Packaging graphics do not originate from aesthetic motives. Their purpose is functional, like protection and consumer function. The functions that package graphics perform are:

1. Identify the product contained as to brand, quantity, and probably ingredients
2. Identify the manufacturer or distributor and his address
3. Instruct the consumer how to open the package and how to use the product
4. State any necessary safety cautions
5. State any other mandatory information required by law or regulatory authority
6. Combine the above into a design that presents the information in a planned layout, with the intent of creating a favorable impression toward the product and its manufacturer in the mind of the user.

In total, then, the one function of package graphics is communication.

The graphic spec for a package consists of two parts — a mechanical layout and a set of color standards. The mechanical layout, or "mechanical," is a black-and-white mockup in the exact size and with the exact detail which will be printed on the package or the packaging component. Every word, punctuation mark, trademark registration symbol, and pictorial detail is in the

FUNCTIONAL AND MARKETING CRITERIA

"mechanical" so it can be checked for correctness, completeness, and compliance with all regulatory requirements. Regardless whether the printing will eventually be done on paper, paperboard, plastic, or metal, the "mechanical" is always on paper and usually carries with it approval signatures of the various department heads responsible for its several aspects — law, marketing, manufacturing, purchasing, packaging, consumer relations, and design. The color specs are indicated by "chips" of solid ink application on paper and overlays of semiopaque tissue on the mechanical, which show where the individual colors are to appear in the finished print.

The original graphic spec as defined above may be replaced later by a proof print, which shows the graphics in finished form and color. The proof, reproduced in several copies, can then be used as a production standard for multisupplier and multiplant operations. Color tolerances are an aspect of quality control which will be discussed in the last four chapters.

2

Manufacturing and Cost Criteria in Specifications

> *... delivery of material to the packing line is the first step in a series of exposures that will cause a certain percentage of defectives, losses, and a downgrading of appearance and function if not taken into account when the spec is written....*

The previous chapter is concerned mainly with the parts of a spec that govern performance of a package after it leaves the packing plant. This chapter considers the structure and purchasing side of the package spec, its delivery to and performance in the packing plant. A small but important list of criteria concerns the people in the packing organization who have to buy and handle the packing materials:

1. Construction and cost
2. Manner and timeliness of delivery
3. Adaptability to packing-line operations
4. Quality

These criteria become the motivation for writing the "in-plant" parts of a packaging materials spec, including, when necessary, a packing process spec.

All of this is very briefly outlined in Figure 1 under II. Construction, and at this point we shall expand the subject.

I. CONSTRUCTION AND COST

The structural part of a spec is the most basic portion and the major determinant of cost. It fixes the amount of raw material needed to make the package and defines for the supplier the conversion steps required to turn the raw material into the package or component. How the construction spec is organized depends on the complexity of the package. For example, a single-wall plain carton for dried beans requires only one dimensioned print, with a statement of the grade and caliper of paperboard to be used. The shipper, a kraft board bundle or a corrugated box, requires a second print and statement of "Mullen test" to define the weight of the facings. If, however, a package is an assembly of several components, it is best to organize the spec also as an assembly of component specs. An example is an aerosol can, which is composed of a can body, bottom end, breast, valve, valve cup, dip tube, release button, overcap, possibly a tamperproof cover, possibly a paper label if not lithographed, and a shipper. Let us suppose that the basic spec is identified by the number 9058, according to a product coding system where 9= paint, 0=flat, 5=1/2 pint, and 8=can.

Then the can, with bottom and breast, would be noted as 9058-1
valve, valve cut, and dip tube   as 9058-2
release button   as 9058-3
overcap   as 9058-4
tamperproof cover   as 9058-5
paper label   as 9058-6
and shipper   as 9058-7

This arrangement assumes that any components which are always constant may be grouped, while those which can be changed individually must be specified individually. In the example above, a 1/2-pint aerosol paint can body will always use

MANUFACTURING AND COST CRITERIA 31

the same bottom and breast spec with a body spec, so those three may be grouped. Likewise, a fixed combination of valve, cup, and dip tube will be used for that size can, but they will all three be some other constant combination for another size of can. Should any part of the spec be changed — for instance, the paper label to a foil label — the basic spec number is retained, and only the last digit need be changed or added to, such as 9058-6 for paper label to 9058-61 for foil label.

In this complex situation, each component requires its own blueprint and statement of raw materials for the structural spec, and an assembly print showing the relation of all parts to one another in a cross-section drawing is very useful to establish the dimensional tolerances permissible in the components.

Many spec writers think of a packaging-materials spec as being satisfactory if it states the raw materials variables, with tolerances, and the part dimensions, also with tolerances, to cover the structural features of the package in question. There are, however, many features of the package which determine how well it will run on the packing line. To include these in the spec it is important that the writer be familiar with the equipment on the line or lines and its idiosyncrasies, such as:

1. Method of feeding empty packages or components to the line.
2. Package shape irregularities and projections in relation to line fit
3. Line impacts and their effect on the packages
4. Line abrasion and its effect on the package or its finish
5. Method of case loading
6. Frequency and complexity of line changes from one size package to another
7. Effect of package on sampling and quality inspection procedures, and vice versa
8. Effect of package on line stoppages or jams and any clean-up complications
9. Tendencies of some components to experience greater shrinkage losses than others in the packing process

Obviously, these items relate to quality and cost problems, and those which are not taken into account when the spec is written will inevitably rise later in the form of technical service projects. Thus, we reach the conclusion that a good packaging structural spec must include consideration of the packing process and the quality control procedures that will apply to the process. It is for this reason that *B. Assembly* is included in Figure 1. The description of the packing process need not be a detailed one in the packing spec, since it will be fully covered in an operations or processing manual, but enough information should be outlined in the packaging materials spec to tie in with the operation. The points above will now be considered individually.

A. Method of Feeding Empty Packages or Components to a Packing Line

There are characteristic practices for most package types:

Metal cans — depalletized row by row and layer by layer onto an in-feed conveyor. An alternate procedure for small-volume use is to have cans delivered in their shippers, from which they are hand-dumped onto the conveyor. Ends are fed from paper-wrapped stacks, labels from stacks.

Glass bottles — somewhat the opposite of cans; most bottles are delivered in their shippers and mechanically removed and placed on the in-feed conveyor. A growing practice is to depalletize from bulk delivery units, mostly used when the filled bottles are loaded into six-packs, etc., for distribution. Closures are sorted from bulk delivery and chute-fed. Labels are fed from stacks.

Plastic packages are delivered according to shape. Thermoforms and injection-moulded tubs and bowls are fed from nested stacks, usually separated by air jets for delivery to a conveyor. Blow-moulded bottles may be sorted from bulk units or emptied from their reshippers. Plastic closures are usually sorted from bulk containers, although some that are large and flat are fed from stacks.

Cartons are almost always erected from stacked knocked-down form. Some, as for cake mixes, are formed from flat blanks or even from roll stock around mandrels. Liners and overwraps are always roll-fed.

MANUFACTURING AND COST CRITERIA

Flexible packages are roll-fed into form/fill/seal equipment, except for preformed bags, which are fed, like cartons, from stacks of KD's.

Whatever the method for the package in question, the delivery of the material to the packing line is the first step in a series of exposures to handling that will cause a certain percentage of defectives, losses, and a downgrading of appearance and function if not taken into account when the spec is written so as to build in adequate resistance.

B. Package Irregularities and Projections in Relation to Line Fit

When any package is designed, the functions it must perform will fix to a large extent its general shape. The functional criteria will be modified by the limitations of the package manufacturing process to eliminate extremes of shape, but there will almost always be some part of the package which is sensitive to abuse on the packing line. Some of these, by package types, are:

Cans — the flanges are most subject to damage, and if badly bent, leakers will result because the double seam will be interrupted when the top end is applied. The problem is more severe with fiber cans, the saving grace being that some products packed in such containers do not need airtight seals. Otherwise, it is not uncommon for a fiber can packing line to have a flange straightener just ahead of the filler or closer to repair flange damage inflicted during palletizing, depalletizing, and intermediate shipping.
The flanges of lithographed cans scratch one another on the body near the open end. This can be minimized, but not eliminated, by tight packing to restrain can movement during delivery, and by can separation with only short surges on the packing line.

Glass bottles — simple squat, cylindrical shapes like baby-food jars, applesauce jars, and small beer bottles are the most trouble-free for packing-line efficiency. Features that cause problems are tall, thin shapes, which topple easily and jam conveyors; nonround cross-sections, which require orientation through the whole line; maximum dimensions above the center of gravity, which requires pocketing on the line to keep from pushing one another over; sharp angles and pinchins, which cause weak spots; dark-colored glass, which is hard to inspect.

Plastic packages — blown bottles although not so subject to damage on the packing line as glass, are more likely to topple and jam because of their light weight. Large shoulders which bump one another or rub guide rails are likely to knock bottles down, as are sudden changes of speed or direction on conveyors. Injection-moulded or thermoformed bowls and tubs cannot be moved on conveyors by line pressure because the top end is the widest part, and they will either fall over or pop upward unless held down by guide rails or pressure plates above the conveyors. The presence of such rails or plates complicates size changeovers.

Closures — plastic closures of irregular shape are the greatest challenges in packaging materials for automatic high-speed feeding. Examples are liquid detergent bottle closures with projecting nozzles, spray nozzles, rollon closures, hinge-top caps, plastic pumps. Chutes to accommodate each type individually are designed for feeding to the capper, but the chief challenge is to "rectify" or upright them from bulk receipt to feed them into the chute.

C. Line Impacts and Their Effect on Packages

The developer of a spec can inspect his packing lines for several conditions which determine in part how strong the packaging materials need be. A few technical rules will help:

1. The damage of impacts varies as the square of the speed. In simple terms, a package may be assumed to be moving on a conveyor at 1 foot per second, and it is stopped by bumping into a surge of packages at the in-feed to some machine on the line. Its momentum while moving was its mass multiplied by its velocity. It lost that momentum at the instant of bumping into the line of packages ahead of it, with a force equal to its mass times the deceleration. The work done in the impact was equal to the kinetic energy in

MANUFACTURING AND COST CRITERIA 35

    the moving package, which was 1/2 the mass times the square of the velocity. If the package weighed 1 pound, the work of stopping it was 1/2 foot-pound, and all this work went into deformation, vibration, sound, and an undetectably small amount of heat. Now, had the package been moving at 2 feet per second at the time of impact, the work done would have been 2 foot-pounds, four times as much as at half the speed.

2. The damage of impacts is proportional to the weight of the package. A 2-pound package will have to absorb twice the work of impact as a 1-pound package, assuming both moving at the same speed. Another way of looking at this law is that empty packages of a given type are not so likely to be damaged as full ones. That part of the packing line which is handling empties will probably not cause as much damage as the parts handling filled packages.

3. Heat impacts add to mechanical stresses. This is particularly important only to the hot-filling or pasteurizing of glass with food and pressurized products. Soda-lime glass for containers of uncomplicated shapes has a heat-shock resistance of 140°F, which is to say that it will stand being filled with products no more than 140° hotter than the glass. If the product is filled at 210°, the glass should be no colder than 70°. Bottles delivered in winter are liable to be colder when taken from a truck or railcar, and if fed directly to a hot-fill line, they must be preheated.
Cold-shock embrittles certain plastic containers, such as polypropylene and clear styrene. For this reason, polypropylene is not usable for frozen foods, and polystyrene must be impact-modified with butadiene to make a satisfactory package for ice creams and sherbets.

4. Liquid products add hydrodynamic shock or "hammer" to impacts on packing lines. Unlike solids, liquids have little internal friction, and when a liquid-filled package is impacted, the full thrust of momentum in the liquid vectors toward the impact.

The most common symptoms of impact damage on packing lines are:

Cans: body dents and open ends out-of-round; dented flanges
Glass containers: breakage, chipped finishes, and scratches leading to breakage
Plastic rigid packages: cracks and chips, clear styrene only
Flexible packages: wrinkles
Cartons: dents, window punctures
Corrugated shippers: dents, punctures, crushed partitions

D. Line Abrasion and Its Effect on Package or Finish

On most types of packages, abrasion causes only appearance defects, such as scratches on smooth surfaces or in decoration. The writer of a spec must recognize the action of his packing lines in respect to abrasion and either arrange for improvements in the lines or specify packages to meet existing conditions. Suggestions for reducing line abrasion include such items as:

1. Plastic rather than metal conveyor plates
2. Nylon, polyformaldehyde, or polycarbonate plastic guide rails
3. Fiber-filled plastic rather than metal star-wheels into and out of line machines
4. Electrical hookups to shut down conveyor belts when the rest of the line is down, rather than continuing to run and abrade stationary packages
5. Line maintenance with periodic inspection to countersink screw heads and remove burrs on parts that could contact and scratch package surfaces
6. Frequent cleanup of glues and other auxiliary materials that can accumulate on conveyors and guide rails and form dry abrasive deposits
7. Slow, steady line speeds rather than high, intermittent speeds

The principal armor against abrasion on carton and flexible package surfaces, paper labels, and can lithography is a hard, high-gloss lacquer applied over the printing. The formulations are all synthetic resins, some being heat or u-v curing varnishes and others being solvent-applied lacquers. The choice among them is based on cost versus performance and the available facilities for

MANUFACTURING AND COST CRITERIA

application at the package supplier's plants. The most abrasion-resistant are considerably more expensive than those that merely add gloss and some abrasion resistance, such that it is cheaper in the long run to modify and maintain packing lines in good condition for low abrasion, rather than pay on a continuing basis for the hardest varnishes or lacquers.

Aluminum cans, glass containers, and flexible packages can suffer more than appearance damage as a result of line abrasion. Double seams of cans made with aluminum ends are quite soft and are readily abraded to the point of cutthrough by steel conveyor belts. As for glass, breakage is rarely the result of a single hard knock on a virgin container. Rather, successive and even mild impacts and abrasion cause scratches and small surface defects that gradually lengthen and deepen with subsequent vibration and shocks, finally penetrating entirely through the container wall and causing failure. Therefore, hardly anything worse could be done to glass containers than to leave them standing in a line on a running conveyor belt over a lunchtime shutdown, vibrating and jiggling against one another.

Flexible packages can suffer seal failure from abrasion, if there occurs a repeated flexing at a corner or the inside edge of a heat seal.

E. Method of Case Loading

Some packages are dropped into their shippers, others are pushed in upward, or endwise, or sidewise mechanically, some are hand-packed. The kind of packing setup will of course have an important effect on the construction of the shipper, second only to the protective needs of the product and unit packages.

The spec writer must bear in mind that the shipper prescribed by Rule 41 of the Uniform Freight Classification is a minimum structure for insurance coverage by the underwriters of the common carriers. This does not assure him that the shipper will be satisfactory for processing in his own plant, and it may be that the packing-plant case-handling system is rougher than average distribution stresses. Many glass packs which for shipping could use nontest partitions must have more expensive test partitions to go through certain automatic uncasers and case loaders, for instance. In another instance, where empty cans are delivered in reshippers, the case dimensions must be larger than otherwise for uncasing, because the flanged open ends are larger in diameter than they will be when later seamed and repacked into the cases.

F. Frequency and Difficulty of Line Changes

The packing-department head of any plant considers himself very fortunate if he can run a "pat" line that never changes specs, and so do the plant engineer and plant manager. It usually takes a whole shift out of production if a high-speed complex line has to be changed from one size to another of cartons or glass packages, and can size changes may take a week. For all the great volume in today's markets for packaged goods, there is more segmentation, with lesser volumes of several package specs to make up a larger total. The spec writer can ease the pains of line changeovers by utilizing interchangeable components for two or more specs in his original design work. He should know the costs in time and money for changes of each component as well as for the whole. For instance, if a line is set up to pack three sizes of liquid shampoo and it is desired to set up for three sizes of a new liquid dishwashing detergent, the components to be covered in any changeover will be the bottle, the cap, the label, and the shipper. For marketing reasons the three detergent bottles will probably have to be larger than the respective shampoo bottles, but if the same caps can be used for the new set, the plant would stay with the same three sizes rather than six. It may be even more troublesome to change the labeler to add three new sizes than to change the cap-feeding setup, in which case it also becomes very desirable to use the three shampoo labels as the blanks for the three new detergent bottles.

G. Effect of Package on Inspection and Quality Control

Determining or estimating the quality level of packaging materials shipments is a cost factor that should be justifiable rather than just a ritual. More will be said about this in Chapter 4, but a few guidelines mentioned here will be useful for orientation:

1. Identify and classify defects and their allowable limits first

MANUFACTURING AND COST CRITERIA

2. Using sampling tables, decide the cost of the sampling and testing program that will yield the desired information on the lots delivered
3. Find out from the supplier how much of this sampling and testing is already done by him — there is no point in duplicating his quality control program
4. For those important variables or attributes that the supplier does not test, conduct the minimum testing that will prevent problems in packing and distribution
5. Where possible, do nondestructive testing, since this procedure results in discarding less material.

As a general principle, it is most efficient to adopt the policy that the packaging materials supplier is responsible for the quality control of his products, while the buyer is responsible for the quality control of his packing operations and packed product. If the two parties have the good rapport to agree on this principle, the packer need only be concerned about the effect on materials quality of the carrier who brought the materials to his plant and the effect of his own packing lines on the materials.

H. Effect of Line Stoppages, Jams, and Cleanup Operations

The layout of a packing line will determine how much trouble a cleanup will cause in the event of a jam or broken package. This, in turn, will help fix the importance of many defects and the dimensional tolerances of packages that will run on the line. For example, can double seaming permits a tolerance of ± 0.030 inch maximum height variation, but only ± 0.005 inch in diameter. If the open end is much out of round as it enters the clincher that hooks the top end onto the body, the end will enter the closer tilted and cause a jam. Should this happen to be a vacuum closer, the line has to be shut down and the closer opened with large wrenches — a job requiring not less than two men and a quarter hour. Obviously, outages of dimensional tolerances and out-of-round tolerances on cans are Class A defects.

On glass lines, cleanup time depends on the nature of the product being filled, where the jam or break occurs, and the accessibility of the site. If a food or beverage is the product, great precautions must be taken to see that no broken glass enters another container. Thus, if the break occurs at a point on the line where the bottles are filled but not closed, it is general practice to discard all the containers between filler and capper. Should the line be laid out in such a way that an adjacent conveyor carries empties from the cleaner to the filler, those also will have to be removed and at least rerouted through the cleaner. Naturally, if the product is a liquid laundry compound, the precautions need not be so rigorous. The moral here is that the spec for the glass container should recognize whatever complications will occur on the line when jams and breaks occur.

There is of course a balancing factor that prevents packing-line conditions from forcing every bottle design into a returnable-weight Boston round, and that is the package cost. The same is true of cans, cartons, and flexible packaging materials: can steel can be specified so heavy that the body will not dent or get out of round, carton board so heavy that it cannot crush on the line, and flexible laminates so strong that bags will not break when stepped on, but the cost will become unaffordable. The best answer is to construct packing lines to handle packaging materials as gently as possible, so that continuing cost penalties are not incurred just to get the materials through the packing plant. A good rule to follow is that the plant-inflicted stresses should be no greater than distribution and field stresses; thus, the spec that suffices to get the product from plant to consumer will be adequate for packing operations.

I. Recognition of Losses from In-Plant Shrinkage

When a packaging spec is written, the materials cost is always calculated, and if it supersedes another spec, a cost comparison is made. At this time a usually neglected factor should be included — differences which may exist in packaging materials shrinkage between old and new specs. Examples of changes that can cause shrinkage increases are:

1. *Glass*: round to square or odd-shaped bottles — nonround shapes break more easily; short to tall shapes — tall containers will topple more readily
2. *Cans*: paper labels to lithography — it is more expensive to discard lithographed cans which are dented on the line before they would otherwise be labeled

MANUFACTURING AND COST CRITERIA

3. *Plastic containers*: paper labels to screened or hot-stamped decoration — same situation as above with cans
4. *Paper to foil labels*: on any of the above — foil labels tend to curl and to double-feed more than paper labels
5. *Preformed bags to form-fill-seal pouches*: the packaging material for the latter is less expensive per thousand, but daily startup and changeover losses are greater because of the need to establish equilibrium running conditions in heat sealers
6. *Closures*: metal to plastic — moulded caps scratch and crack more readily than metal caps

In purchasing, it will be necessary to order a few percent more of those components which have greater in-plant shrinkage. Other than those items mentioned above, shipping containers will always be needed in excess of the amount necessary to match the unit packages. Corrugated boxes, the major package type in this category, always suffer loss from crushing in uncasers and casers, besides being convenient for carrying samples for quality control, research, and sales internal uses.

This concludes the discussion of the nine points related to packing-line considerations which affect package construction and cost. They are just that — considerations — for study by the spec writer at the time of development, not to be included in the spec, which would only add to its bulk, but to be evaluated so a balance can be reached among the cost of packing-line improvements, and possible package cost penalties, and technical service problems.

What follows in this chapter deals with the other three broad spec criteria of interest to the people in the packing company.

II. MANNER AND TIMELINESS OF DELIVERY

Suppliers of packaging materials may be located at any distance from the packing operation, as close as next door or halfway around the earth. The smooth operation of a packing line depends, of course, on a supply of packages or components delivered soon enough to be inspected and brought to the line, and in such form and packing as to be fed continuously to the packing line.

Timeliness of delivery is always important; although the manner of packing for inbound shipment is not very important to a slow or largely hand-operated line, it is vital to the output of a high-speed line. It is in such a line that a great investment has been made for the express purpose of high output, and to frustrate that purpose is to waste the investment. What speed is high speed? A convenient separation between high and low speed is 100 packages per minute, based on the following analysis:

Assume that a packing line is filling or closing metal cans or glass bottles. The packages are delivered in their shipping cases, with top ends of the cases unglued. Pallet loads of the cases are brought by fork truck to the front end of the packing line, which starts with a conveyor leading to a washer. One man can take cases off the pallets and remove cans with both hands and place them on the conveyor at 100 per minute. Once he has opened the case, he can remove four cans at a time and transfer them at about one motion per second to the conveyor. This is nominally 240 cans per minute, but he will lose time putting empty cases aside and opening full ones. Naturally, he cannot keep on at the nominal rate for 8 hours and will need a relief man for breaks and lunch. Above this rate of about 100 per minute, a second full-time operator would be needed, and the added cost would justify the installation of uncasing equipment, to which the single operator would merely feed full cases of packages. If the average case holds 24 packages, he has to handle only 8.3 cases per minute to feed 200 cans or bottles per minute.

In a multiplant operation, it is often necessary to leave it to the plant to deal with the local supplier on the manner of packing for delivery, rather than building this detail into the spec. Both the supplier's plant and the packing plant may have facilities and layout such that there are advantages to receiving in a certain way that are not common to other plants. For example, the supplier, although distant, may have a local warehouse in which he can inventory materials for the customer to pick up by his own trucks or a local transfer service. Or, in another case, he may park trailer-loads of packaging materials in the customer's yard, with a driver to shuttle them to the unloading docks on a few minutes' notice. In either case, palletized delivery would be advantageous. If the customer uses small quantities intermittently, pallets would be an unnecessary cost.

MANUFACTURING AND COST CRITERIA 43

Plastic containers are commonly shipped to packers either in bulk or in nested reshippers. A large customer may very well have one plant equipped to receive in bulk, while another must receive in reshippers. The former would be cheaper where feasible, and the cost difference could justify a bulk sorting installation at the plant not so equipped.

The point is, in any case, that the basic spec becomes confusing if it includes the details of packing for delivery to both plants and their respective suppliers. If both of the packer's plants are supplied by one or two plants of the same supplier, the purchase contract can reflect the differential between the two methods of packing for delivery. This will be considered further in the next chapter.

III. QUALITY

The criteria for adaptability of a package spec to packing-line operations have been adequately discussed above, so we may proceed directly to the matter of quality. From the packer's viewpoint, the intent of a statement on quality of packaging materials is that they operate well on his packing lines, protect his products, and perform their consumer functions. Within the context of this chapter, which is concerned with in-plant performance, the criteria for quality will include the following factors:

1. Acceptable raw material, with acceptable limits of some kind; e.g., can plate of certain gauge, temper, and plating; or polyethylene of certain grade (food, film, injection-moulding, or other), color, and melt index; or aluminum foil of certain gauge, temper, and limiting pinhole count
2. Formed or fabricated dimensions, with dimensional limits, usually on prints
3. Performance requirements with limits on certain attributes, such as damage resistance on packing lines
4. List of defects, classified as to critical, major, and minor, with limits
5. Sampling and inspection schedules and procedures, where appropriate
6. Statement on action to be taken in case of nonconformance to spec by a given lot of delivered materials

7. Color standards, where applicable, usually with light and dark tolerance limits, and graphic defects classified as above for physical defects.

At the time of establishing the tolerances for quality, it must be recognized that there will inevitably be occasions when a shipment of delivered materials will contain defective elements, which, stated another way, means that some material or packages will be outside the established tolerances. This raises more than one question — How do you determine how many defectives there are in the lot? How many can be tolerated? What are the implications of rejecting the lot? Of sorting out the defectives?

The answers are not simple and depend on the severity as well as the frequency of the defects and also on the urgency to use the material. But these answers depend on good communications between supplier and buyer, set up before rather than after the discovery of problems. After all, if defectives were not expected, the whole matter of building quality criteria and limits into specs would be nothing more than a technical exercise.

This bring us, then, to the questions of how a supplier of packaging material relates to his customer, the packer; who gets involved in the relationship on both sides; what their communications problems are; and how they may be solved.

We have examined the broad format of specifications, what they should include to satisfy the packer's marketing requirements and his manufacturing operations. Chapters 5 through 8 will go into specifications and quality control of specific packaging-materials types in detail.

3
The Buyer-Supplier Relationship

> *Let us now shift our attention from the communication itself and focus on the persons who participate in it.... In today's complex world, a supplier is many people, and a buyer is many people.*

In this light of what has been discussed, the packaging specification has hopefully emerged as a communication rather than a document. The spec communicates factual data and intentions on package structure and function, and quality standards which determine acceptability. Let us now shift our viewpoint from the communication itself and focus on the persons who participate in the communication. It was stated on the first page that the objective of all this communication is true understanding of what is wanted and what can be tolerated. This, of course, refers to people, and in today's complex world, a supplier is many people and a buyer is many people.

We need next to recognize that there are more communications involved that just the spec and the purchase order to supply the material, and that the many parties concerned must be "wired in" to the communications.

It is not the purpose of this chapter or book to serve as a text on purchasing itself as an art and science, but rather to organize our thinking about the communicative environment in which purchasing is done. We shall therefore assume that the packaging-materials buyer knows his job, and that suppliers' representatives with whom he deals are competent in their sales function. In the case of any given package, the two, buyer and seller, are armed with two documents, the spec and the purchase contract. It is now largely up to the other persons in the drama to play the leads in the next acts. (See Figure 2.)

A safe assumption may be made that several background conditions will have changed since the development of the spec was started, and that changes will continue to take place; for instance:

1. The technical man who worked on the spec development at the supplier's company or the user's company, or both, will be transferred or promoted to another job, with resultant loss of their experience.
2. The plant originally scheduled to run the new spec will be changed to another plant, possibly requiring a change also of supplier plant.
3. Some of the fabricating equipment at the supplier plant will be modified, or some of the packing-line machinery at the packing plant will be altered or replaced, with changes in performance one way or the other.
4. The plant people at either end who approved and/or tested the original spec will be moved.

What else could happen that would be nonroutine? While the viewpoint is on negatives, everything may as well be put on the table:

5. The spec doesn't tell everything. We noted at the outset that the spec is the buyer's, not the seller's, and the latter will have more details in his own spec to cover his fabricating process. For instance, a carton-maker may be supplying a candy box, in which the glue he uses for fabrication is not specified by the customer. He may feel free to change to a glue which runs better on his machinery or which costs less, but if the

```
┌─────────────────────────────────────────────────────────────────┐
│                                                                 │
│            XYZ COMPANY        1000 WEST 388 ST.                 │
│                               NEW YORK, N.Y. 90099              │
│         PURCHASE ORDER     NO. __15750__   DATE __10 Feb 1985__ │
│                                                                 │
│     SUPPLIER                   CONSIGNEE                        │
│         Rainbow Label Co.         XYZ Company, Plant 2          │
│         100 Impress St.           Forest, Me.                   │
│         Gutenberg, Vt.                                          │
│─────────────────────────────────────────────────────────────────│
│     AFTER DELIVERY, SEND DUPLICATE INVOICE AND RECEIPT FROM CONSIGNEE TO: │
│                    ACCOUNTS PAYABLE, ABOVE ADDRESS              │
│─────────────────────────────────────────────────────────────────│
│     WHEN WANTED _third week March, 1985_ SHIP VIA _truck, supplier's choice_ │
│     QUANTITY      DESCRIPTION / SPECIFICATION            PRICE  │
│                                                                 │
│     500,000       Spec # 9058-6, labels for ½ pint aerosol  $ 4.10 │
│                   paint can.  Lot to be composed of 5 items:   per │
│                                                               1000 │
│                   Minimum Quantity    Design #    Color        │
│                       100,000           3500      White        │
│                       100,000           3501      Red          │
│                       100,000           3502      Yellow       │
│                       100,000           3503      Blue         │
│                       100,000           3504      Black        │
│                                                                 │
│                   Labels to be banded in stacks of 500, 10 stacks/carton, │
│                   4 cartons per case.  Tolerances: no underage, 5% overage. │
│                   Design # and quantity to be stencilled on cases. │
│                   No more than one Design # in any case.       │
│                                                                 │
│                                                                 │
│     COPY 1. SUPPLIER    CONFIRM OR ADVISE IMMEDIATELY IF UNABLE │
│          2 & 3. CONSIGNEE   TO MEET THE REQUIREMENTS ABOVE.     │
│          4. ACCTS PAY.                                          │
│          5. ORIGINATOR                                          │
│          6. PURCH. FILE  BY _A. N. Johnston_ A. N. Johnston , BUYER │
│                                                                 │
└─────────────────────────────────────────────────────────────────┘
```

Figure 2 Typical purchase order for packaging material.

new glue incidentally imparts an odor to the boxes, the customer could reject them. Or a glass-maker could unilaterally change the coating on his bottles, only to find that his customer's labels won't stick with the new coating.
6. The packer could install a process change that makes the packing material obsolete. For example, the cans which ran perfectly on a 250-per-minute line may scratch and dent badly on a 500-a-minute new line, or even in a new case packer installed on the old line. A change from batch to continuous retorting could have the same effect on glass jars, or again, a new uncaser and case packer.
7. Strikes and threatened strikes add a note of urgency to production which tends to relax attention on quality. On the personnel side, overtime demands may be heavy, and normal alertness lags; crews other than those experienced with a given spec may have to run it; at the end of a strike experienced hands may have gone elsewhere, leaving the less experienced to start up in a period of heavy demand. On the materials side, when a stockpile is needed, one of the raw materials may be delivered out of spec. Should it be used? Suppose it is a paper/foil/polyethylene laminate and the paper is 5% below minimum specified weight and strength. The convertor could compensate for the weight shortage by increasing the weight of the glue laminant correspondingly, but will the laminate perform, or will the added glue make it too stiff to form and fill, and will its stiffness cause the foil to pinhole? If the light weight is not compensated, will the laminate be strong enough for machine stresses?

On the equipment side, out-of-control process conditions may be accepted in times of labor unrest. An extruder applying a heat-seal coating of polyethylene below minimum thickness may produce a laminate that will not seal properly, or one whose seals are too thin to prevent leakage when contaminated with product particles. A solvent coater may cause equally bad problems when running over specified wet thickness: the coating may run before drying and become uneven, the coating may "skin" and bubble in the dryer, making nonsealing spots, solvent may be retained and cause off-odor, and heavy coatings could build up on heat-seal surfaces in packing.

THE BUYER-SUPPLIER RELATIONSHIP 49

Add to the above all the routine possibilities for trouble, such as breakdowns, out-of-spec raw materials when packages are to be fabricated on a close schedule, normal turnover of personnel, change of computer hardware or software in production scheduling and inventory control, and it almost seems remarkable that a spec is ever met. That it is met most of the time is a testimony to the competence and communication skills of the responsible people. Let us look at the communications which take place in a normal shipment, when specs and a purchase order have been established, and a packing plant releases a piece of the total contracted packaging materials from a local supplier plant. See Figure 3.

How many of the people in this grid should have the customer's spec? The key people are the Quality Control Supervisors in both the supplier and customer plants. Everyone else in the grid, particularly the Production Department Heads, presumably accepted and committed to meet it when it was first presented to them; by their sales staff in the case of the supplier's people, and by their packaging manager on the customer's side.

The first observation to be made on Figure 3 is that there are a lot of people involved, even when the activity is routine. Other than the clerks, all of them are responsible to use initiative and judgment in reaching the decisions which they communicate. These judgments relate to lead time, allowance for shrinkage, acceptance or rejection, overtime, manpower, and cost. Questions that arise often lead to more communication, which hopefully will have the purpose and result of resolving into mutually satisfactory action. For example, the two Quality Control Supervisors should have such rapport that the supplier's man can call the customer's man if he finds his paper below spec, and ask if he can tolerate it as is, or whether the schedule for finished goods permits a delay to get a new lot of raw material. The question of available time can be resolved with the customer's Production Control Supervisor. If time permits, the lightweight paper can be set aside for some other conversion order whose spec it meets, and there is no loss. Should time not permit, the customer is aware of the quality hazard and can make every possible adjustment to accommodate to the situation, such as run the laminate at slower speed, or decrease the tension on the sheet as it runs through the

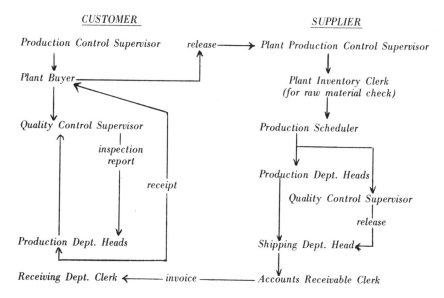

Figure 3 Communications flow in a routine order against contract for packaging materials.

the packing line. This kind of situation, if handled with a constructive attitude, can lead to a valuable cost reduction by learning to run a lighter material than was previously considered acceptable.

What other communications take place between the customer and supplier plants? Several items normally come up for consideration:

1. Overages and underages on a given order. The customer's Production Control Supervisor will have an order from his headquarters for a given number of cases of finished goods shipped. He needs first to know from his Production Department heads and his Receiving and Shipping Department how much extra to order to cover losses from damage, sampling, and inspection, and he must add this quantity to his packaging-materials requirement. Next he must check with his counterpart at the supplier plant what his underage might be on the order in question, and add again to correct for that factor. Example: The customer plant is required to ship 5000 cases of electric clocks, 10 per case, each in its own carton. How many cartons shall he order? The number shipped must be 50,000. The Production Department advises that he should allow 3% losses in-plant, so he must add 1500 to the order. He then tells his own buyer that he must have 51,500 cartons delivered, and the buyer so advises his contact at the supplier plant. The carton-manufacturing rules of the house quote plus or minus 10% on an order of this size, so if their own Production Control Supervisor puts through the order at 51,500 cartons, they may be over or under that number by 5150. Obviously, the lower quantity would be unacceptable, so the order must be put through at 10% over 51,500, or 56,650 cartons. The customer may receive that number, or as few as 51,500, or as many as 61,800 cartons. Assuming tht the cartons cost $0.105 each, the customer is risking an excess cost for 10,300 unwanted cartons, or $1081 to ensure himself of getting his required 51,500 cartons. If this box spec is one which is used routinely, say every quarter, the overage is unimportant, since it can be inventoried and deducted from the next-quarter requirement. This, of course, is not without risk. Leftovers are not usually given the best of treatment, and small quantities of inventoried KD cartons need to be stored in a clean dry space in order to be usable when wanted, and an accurate, available record of their existence needs to be made available to someone who can use it—another communications problem.

If the carton in question is not used routinely, as, for instance, a special holiday package, any overage is a net loss. Here the supplier and customer may be able to save a few hundred dollars by close communication with the supplier's Production Department, and the normal variability of 10% around the nominal order might be narrowed.

2. *Manner of packing for delivery to the customer.* In making a basic quotation, most suppliers assume a standard packing method that they regularly use—KD corrugated boxes tied with cord, rolls of laminate in film wraps and fiber drums or corrugated boxes, plastic bottles in bulk pack, metal cans palletized, etc. There may be other arrangements advantageous to any given pair of supplier-customer plants, and in order to provide for such options, many customers' specs do not specify the manner of pack-

ing for inbound shipment. Alternative packs may cost a bit more to the customer, but may provide some specific convenience, such as labor-saving in unpacking, or reusability of the outer packing, or reduction of waste material.

3. Storage of packaging materials. There will be occasions when the customer will have ordered packaging materials, and his headquarters postpones the use of those materials to interject more urgent production orders. When this happens after the supplier has already made the materials for the postponed order, an agreement must be made for holding the materials, by the supplier, or the customer, or at an outside warehouse. Such arrangements will usually involve the supplier's sales department, since the payment for the packaging material produced will become part of the agreement, and also payment for the storage of it. Most users of packaging materials do not provide large spaces for the storage thereof, so the choice is usually in favor of the supplier's warehouse or outside storage for which the customer pays.

The communication in this kind of situation is initiated by the customer's Plant Buyer, who advises the supplier contact of the postponement, with an estimate of its duration. The supplier Plant Manager will be operating under his own management rules, which include limitations on how much and for how long he may store customer materials, and whether and when he should invoice the customer for such materials. If the customer's request falls within his given limitations and available storage space, the supplier and customer can make the agreement at the plant level, but if not, the supplier sales staff will be contacted and will possibly meet with customer headquarters staff and management to make firm the terms of an agreement to handle the situation.

4. Arrangement for plant vacation and shutdown schedules. When a customer plant closes for vacation, materials coming routinely from a supplier must be cut off for the duration of the shutdown. This is the simpler of the two situations, for when the supplier plant shuts down, he must make enough materials ahead to keep his customer running, not only during the shutdown, but also for as long as it takes him to resume deliveries after startup at the end of the vacation.

5. Supply for peak production periods at the packer's plant. Few items are packed year 'round at a uniform rate, and if the

THE BUYER-SUPPLIER RELATIONSHIP

customer's peak rate exceeds the supplier's converting capacity, he will need lead time to produce packaging materials ahead of his customer. The following example illustrates such a situation: A packer of orange juice uses glass containers. He packs 15,000 cases per week except for the months of November through February, in which he packs 30,000 cases a week. His glass supplier can produce at a maximum rate of 25,000 cases per week for the juice packer. How long ahead of November 1st must he start to stockpile glass for the peak production requirement? The peak period consists of 18 weeks, during which the packer runs ahead of the supplier by 5000 cases a week. The total stockpile must therefore be 18 × 5000, or 90,000 cases. Before the peak period the supplier can exceed the packer's needs by 10,000 cases per week. He must start to run at his peak rate of 25,000 cases per week 9 weeks before the start of the peak packing period to stockpile 90,000 cases of glass—about September 1. In order for this action to be taken, the communication must be made some time ahead of September 1, and it must be initiated by the packer's Production Control Supervisor to his own Plant Buyer in order that the latter may make the proper plan with the supplier.

It is worth noting that the price of communication lapse in this case would be quite high—insufficient glass to pack in the peak period. The supplier is limited by the number of moulds he has to make the glass required, and if he does not get word to run at peak production starting September 1, there is not time to make an additional set of moulds and get them in operation by November, even if his production schedule has open time.

Good planning to provide lead time can minimize the cost of tooling, such as glass or plastic moulds. In the case above, one set of moulds will supply the packer's needs, given communications that allow their most efficient use. Otherwise, a second set of moulds will cost about $8000.

So much for the routine events on which packaging materials suppliers and their customers communicate. Let us now look at some of the more developmental aspects of business in which the two parties can fruitfully cooperate. In these matters the question of security arises: a supplier who develops a new packaging material or process for making a new material can take steps to apply for a patent before revealing it to a customer. He cannot prove its value, however, or gain commercial value from his invention unless he sells the new product to a customer. In the first production there will usually be revealed some improvements which add to the original invention, and in which the customer participates. By virtue of this participation, the customer gains the knowledge of using the new material before anyone else, in return for his expenditure of time and effort. The customer could have avoided the expense of participation by declining the offer to try the new material, in effect by saying, "Develop the new material fully to a commercial state and tell us when it's ready." His risk is that it might be good for his business to use the new material, but it may appear first as a competitor's package.

The moral is that the customer can make an educated judgment of the value of the new material if his development staff are on an open channel of communication with their counterparts in the supplier company. In this case, where the supplier reveals his invention to the customer, the latter has no security risk, but it would be interesting to look at the opposite situation.

Suppose the customer conceives a new and improved package, which he wishes to have produced for himself exclusively. His first step is to apply for patent coverage and then ask his supplier to make it for him. It is rarely, however, that an invention conceived on paper or in the laboratory can become commercial without manufacturing tests which reveal the need for and lead to further inventions and improvements. In this case the supplier is more expert than the customer and more likely to come up with the innovative improvements. Now who owns the commercially valuable invention? The customer has assumed a security risk, not in revealing the basic invention, which he took steps to protect, but in the probability that his invention is not commercial until the supplier applies his own innovative development skills to the problem. Having done so, the supplier is within his rights to apply for patent protection on his own behalf, and indeed he must do so, since patent law permits issuance of a patent only to the true inventor. A patent may be assigned to individuals or an organization other than the inventor, but may not be issued to a delegate. Furthermore, if the supplier does not apply, he runs the risk that a competitor may preempt him. How is this joint venture clarified to the satisfaction of supplier and customer? It must be assumed

THE BUYER-SUPPLIER RELATIONSHIP 55

at the outset that both customer and supplier desire successful commercial use of the new package in question; therefore, there is no point in distrusting one another to the point of blocking that desirable objective. Neither party can "go it alone"; the customer needs the new package—he originally proposed it—and the supplier has no use for it without a customer to pack it. Once started together on the development of the new package, they may cross-license each other on the respective inventions which they contribute, or the supplier can assign his inventions to the customer, since the latter initiated the project. This kind of project obviously involves communications not only between development staffs, but also between legal and patent departments and managements.

What has just been discussed concerns the supplier as an aid to innovation, with consideration of the security angle. Another subject for communication between the parties is the security of supply from interruption for any reason. We spoke above about planning for stockpile production in anticipation of vacation closings and periods of labor unrest. The customer also needs to be aware of other hazards, such as interruption of the supplier's raw materials, fire, flood, or other disasters at his plant or warehouse. Such facts should be known as: the supplier's labor contract expiration dates; the structure of his buildings and the fire protection given them—both control systems, such as sprinklers, and insurance coverage; whether finished packaging materials are stored safe from plant hazards and climatic damage; storage conditions of customer's moulds, dies, and printing plates or cylinders between production runs; excess capacity available in case of unusually large demand for materials; availability of production from alternate plants in the event of a stoppage from the usual point of shipment. In the course of normal communications, these points will be brought up during periodic visits by the two parties to each others' facilities; it is basically the responsibility of the customer's p.a. and buyers to acquire information related to the security of their supplies.

4

Quality-Control Principles

Inspection made after a lot of anything is completed can only be a quality audit—it is too late to exercise control.

Given that good specifications for a package are written and the arrangements for its purchase are successfully concluded, it remains to be demonstrated that numbers of the package in question can be delivered to the packer as close to the spec as he intended them to be. All the activities concerned with manipulating production at the supplier's plant and protecting the packages in transit to the customer are called *quality control*, including:

1. Inspection
2. By appropriate test methods
3. Of representative samples
4. Against established standards
5. For adjustment of manufacturing processes

6. To maintain production at a maximum percent of acceptable units
7. As measured against product specifications

The need for organizing and conducting such a complex of activities lies in man's discovery of the great natural phenomenon, perhaps a natural law, that no two of anything are exactly alike. This applies to man-made articles such as cars, toothbrushes, and glass bottles as well as things of nature, like rabbits, clouds, and mountains. Without action on all of the seven points above, there can be no quality control. When a package is specified, both the buyer and the supplier should realize that no two packages delivered will be precisely identical. The quality control they exercise over the package manufacturing and the packing operations is intended to assure, as much as humanly possible, that the differences which will exist are not important to filling, protecting the product, and presenting it properly to the consumer.

This chapter is devoted to the explanation in terms suited to the packaging specialist of the principles behind the components, listed above, that make up a complete quality-control program. We shall not attempt to expound the mathematics and statistics and deeper theory behind quality control—this is done much more competently in many books by experts on the subject.

I. INSPECTION

Inspection of what? What is *inspection*? The original Latin for *inspect* means "to look into." So *inspection* means "the act of looking into something." To the packaging professional this means looking into anything that affects the quality of packaging materials and packages all the way from the raw material to the packaged product at the point of delivery into the consumer's hands. Inspection, then, is a necessary part of quality evaluation; it is the sensing process for quality aspects. For example, in a modern bakery loaves of bread enter a continuous oven on a conveyor, with the intention that they exit at the other end properly baked. If by chance they come through overdone and "burnt," their condition can be detected by sight, smell, and touch. Three aspects of quality have thus been "inspected" by three organoleptic senses. Any one of them alone would have yielded the same conclusion. Inspection by sight of two other indirectly related "controls" might have indicated the likelihood of this undesirable quality occurrence: an oven thermometer and the conveyor-speed gauge. It is most important to note that in this situation inspection per se is not a true control—it is an evaluation after the fact which can do no more than communicate the need for action to establish control, or to say that quality is acceptable and that no action need be taken (when the bread exits from the oven with desirable color, odor, and texture).

What is inspected? Raw materials, converted materials, finished materials, and the processes of conversion, packing, and distribution are all proper areas for inspection. It must be recognized here too, however, that in every case inspection is an evaluation of past action. A raw material has been made, and the inspection is to determine whether it was made properly. A process has been started and is inspected to see whether it is "in control." It is basic to the nature of inspection that, if nothing has been done or made, there is nothing to inspect. Inspection yields information, then, on which to decide whether action is or is not necessary.

The next significant point to be made is that inspection can be direct or indirect. Package technology is essentially a materials science, and therefore any inspection of a *process* is an indirect inspection. It is assumed, for instance, on the basis of much experience, that annealing temperature is important to the physical properties of glass bottles or can-making aluminum or steel. Annealing temperature is therefore carefully inspected to make decisions on control action. Inspection of the physical properties in question, on the other hand, is direct inspection. The indirect inspection is used because the information it yields can be put into action sooner than that based on inspection of packages, and because indirect inspection is nondestructive. The strength of glass bottles can only be inspected by testing bottles to breakage, and the stiffness of can metal can be inspected only by cutting and testing samples. In an economic sense, the destruction of a part of production is wasteful whenever indirect inspection is an alternative.

A final point is that the agency which inspects may be human or instrumental, but whichever it is, the information it communicates must be used by another agency to act in relation to control. That latter agency may also be either human or instrumental. To consider again the case of the bread baking in the continuous oven, it can be arranged that the conveyor speed be fixed and invariable, and that the inspection be done by an oven thermometer which adjusts a fuel valve or air intake, via powered valves and dampers which respond to electrical signals from the thermometer.

We are now ready to summarize the part that inspection plays in a quality-control program for packaging, or anything else, for that matter: inspection is the communication of an existing situation, either in a lot of material or in a process, as it relates to one or more quality aspects of the material.

II. APPROPRIATE TEST METHODS

We have spoken about inspection of quality aspects in packaging materials and manufacturing processes. It is essential to know whether the aspects inspected actually relate to the quality in question, and how accurately. As in other fields, the packaging specialist, too, can be influenced to accept certain measures as meaningful because he has heard them often used and assumes that they are generally proven to be true. An outstanding example of this situation is the common acceptance of the Mullen test for corrugated board as a measure of the protection that will be given products contained in a box made therefrom. The Mullen test measures the bursting strength of corrugated or other board, and it clearly shows that as board weight is increased the "Mullen test" increases. The leadership of two national technical associations of packaging and materials-handling experts became convinced, however, that Mullen or bursting strength is not an accurate indicator of protective level. They had reason to believe, furthermore, that overdependence on Mullen tests is the root of much freight damage and monetary loss, and they have developed, via ASTM Committee D-10, an improved protocol for performance testing packed goods.

What can the packager do to test the protective value of corrugated containers? He can make laboratory abuse tests on finished cases—such as exposure to a vibratory table, followed by impact and/or shock tests or tumbler tests. A shipping test to some field location and return is also a favorite method, but so lacking in

QUALITY-CONTROL PRINCIPLES

control and knowledge of actual exposure conditions that significance can be read into the results only after a series of shipments. In short, the best advice is that if one wishes to measure the damage resistance of a package specification, he should expose samples of the finished pack to a controlled series of handling tests.

One other example will serve to illustrate this matter of relating a test to the quality it is intended to measure. This one concerns the packaging of products in an inert atmosphere, such as nitrogen or carbon dioxide. The package barrier must therefore have resistance to the passage of oxygen. Although not so common today, the testing of barrier materials for selection and quality inspection was done exclusively on flat sheet material, mostly because this was the only test available. In fact, the effects of forming the package, with its folds, creases, seals, and the abrasion of the forming process, usually have a greater effect on the oxygen barrier quality of the finished package than the variability of the basic flat material. Again, the best way to test oxygen permeability of gas packs is to evaluate finished packages.

Many fine technical minds have worked on test methods for packaging materials and packages, and in committee they have correlated interlaboratory tests to determine the accuracy and precision of these tests, and what they really measure. The professional packaging man has access to these methods through such technical associations as ASTM, TAPPI, Fiber Box Association, National Canners Association, The Packaging Institute, the Society of Packaging and Handling Engineers, the Glass Packaging Institute, and the Flexible Packaging Association. Availability of the test methods may be through publications or committees of these associations, whose addresses are listed in the appended references.

Tests must relate in degree as well as in kind to the quality being inspected. The weight of steel plate for can making can be accurately determined, for example, by measuring its thickness, since its density is very constant. But that thickness measurement must be very precise, because an error of one-half of a thousandth of an inch will throw the numerical result off by 4.5 pounds per base box. This is a 4.5% error on a nominal 100-pound plate, but an 8% error in measuring a nominal 55-pound plate. There would be no use in providing a quality inspector of this plate with a thickness gauge which can measure to a precision of 0.002 inch. He would read most of the material he inspects as out of control and needs a gauge that can read to 0.0002 inch to do an effective job.

There is usually a difference between tests which are made on packaging materials and packages while a specification is being developed, as compared to tests which are used for routine quality inspection. When developing a spec, the tests are used to select materials or packages; and a given test may be used only a few times. Its purpose is to evaluate the average differences among materials or packages, in that situation. When testing is done for quality inspection, however, the action is repeated often, and its purpose is to evaluate a range among samples within a given material or package. Some pains are therefore taken to set up the quality inspection tests with permanent equipment in such a way that technician time is minimized. For example, let us consider the development of a spec for baby-food jars that are to use a new, convenient closure. The tests involved would measure vacuum level and opening torque. During development at least two closure types would be compared in sample lot sizes, using general laboratory equipment for the purpose. But when one is specified and production starts, the q.c. lab set up a jig which instantly centers the fixed-size jar for the vacuum test, with a torque tester right beside it, so that a technician can make both tests in a few seconds. The acceptable test limits are preferably marked directly on the scales of the instruments. The apparatus and its arrangement are not suited to any other jar and cap sizes, but are ideally designed for efficient inspection of the spec in use.

A caution must be mentioned in connection with setting up "short-cut" rapid quality tests; this is that in the translation from the original selection tests, something may have been lost, and the accuracy of the test may have been thrown off. In the case of the baby-food package, the vacuum gauge may take 1 second to come to an equilibrium. There was no hurry during the spec development, but if in production the q.c. technician reads the indicator only 1/2 second after puncturing the cap, his results will be inaccurate. The same would happen if in building the q.c. apparatus the vacuum probe were longer than that of the research instru-

QUALITY-CONTROL PRINCIPLES 63

ment. The air in the probe, rushing into the void space under the punctured cap, would result in a lower vacuum reading than with a shorter probe containing less air.

For reasons like this it is important to cross-check new testing apparatus with the original research equipment or other "standard" instrument, if one exists. In the event of finding out-of-limits production samples, the q.c. manager may well be challenged as to whether his testing equipment is working properly, either by the production department in his own plant, or by the packaging-materials supplier, if he reports a fault therein. The usual next step is the "interlaboratory correlation test," but it is too late to think about it when a lot of production is embargoed and pressure is on to identify the defect or release the lot. Most correlation tests are run by committees of technical associations, and it is there that the packaging manager can usually find others with a common interest who would be willing to run one or more "round robin" tests, in which each participant gets for testing some samples from all the other task-force members, following which they meet to compare their results. Presumably they adjust or modify their respective q.c. apparatus and retest until they all get comparable results on an equivalent set of samples. Finally the committee writes and the technical association publishes a "Standard Test Method," with instructions on setup and calibration of the apparatus.

Should there not be others wishing to standardize a given test method, or should the using company not wish to divulge its test methods, the q.c. managers in two or more plants can exchange samples for standardizing or cross-checking their methods, or run audits that use apparatus of the research department as a referee.

III. REPRESENTATIVE SAMPLES

At this point we are getting into deeper waters. We wish to inspect lots of packaging materials, and we are satisifed that the test methods to be used do in fact evaluate important quality aspects. Now, how do we approach the lots of materials to inspect them? Suppose they consist of a million cans, or a million plastic screw caps, or a million repeats on a flexible laminate in roll form. To inspect the entire lot of a million is out of the question, either from the standpoint of time or manpower; or if any of the tests are destructive, there would be nothing left to use. The answer, of course, is to inspect a sampling of the lots, and the difficult part is to select the sample so that it reliably represents the lot in question.

Consider how the three different materials are made:

1. A million *cans* of one spec are most likely made on a single can line, all in sequence over a period of about 40 hours at 400 to 500 cans per minute. They are palletized sequentially, perhaps a thousand to a pallet, and the pallets are numbered to identify date and shift of manufacture. If so ordered, the pallets can be numbered sequentially. The lot 1 million cans will then consist of 1000 pallets.
2. The million plastic caps would be made on a group of injection moulding machines. Let it be assumed for simplicity that the end use permits linerless caps. An average mould would have 12 cavities and a 10-second moulding cycle. Each mould and press thus produces 72 caps per minute, 4320 per hour, and 95,040 in a 22-hour day. The 1 million caps can be moulded in about 11 days with one press. Assuming that the caps are bulk-packed 500 to a corrugated case, the lot consists of 2000 boxes. Code dates on the boxes will show 33 different day/shift combinations.
3. Flexible laminates are generally made about 40 inches wide and slit to finished width. Printing is usually the last operation before slitting. A series of laminators would be used to form the laminate before printing. If the repeats are 5 inches long, and three widths can be slit out of the master roll, the linear length of the laminate is $5{,}000{,}000 \div 3$ inches = 140,000 feet. The lot would all certainly be printed on one machine, let us say gravure, which runs at about 1000 feet per minute. Thus, the final run would take only 140 minutes or about 2 1/2 hours. If each roll consists of 1000 repeats (416 feet), the lot will be made up of 1000 rolls slit from 333 master rolls. There will be only one production date and one shift.

Now, would we sample these three different lots in the same way to pick up a truly representative cross-section? If anything

QUALITY-CONTROL PRINCIPLES 65

went out of control in the manufacture of the three lots, would it be likely to follow the same pattern in all? Let us consider the occurrence of situations that cause defects which could arise in each kind of manufacturing operation.

First, in can-making, let us assume that something happens to the solder bath for body seaming as the can bodies are formed and passed along the body-maker. As a result of this failure, 1000 cans are made all in one sequence with incomplete soldering—they will be leakers. Let us say this happens from the 501,501st can to the 502,500th can. These cans will be on the bottom half of the 502nd pallet and the top half of the 503rd pallet. At 400 per minute, the defective cans will all have been made in 2 1/2 minutes of the production and will comprise 0.1% of the lot of 1 million. Typical of a sequential process is the probability that the defectives will be grouped together in one part of the lot. The sampling plan must be set up in such a way as to pick up some indication of these 1000 defectives, so that the need for a more detailed sampling on that part of the lot is disclosed.

Next, in plastic cap moulding, or any other injection moulding, a common failure is the partial blocking of a gate to one cavity in the mold. This leads to an "unfilled cavity" or "short shot"; the parts coming from the blocked cavity are incomplete. In the 12-cavity mould of our example, this means that 1/12 of the caps produced during such a period of blockage will be defective. Should this condition persist for 4 hours of production, due to an inexperienced operator, 1440 bad caps will be produced. These defectives will be randomly mixed with 12 times their number, or 17,680 caps. Somewhere in the total lot of 1 million caps there will be 36 boxes containing 1/12 defectives. Will the same sampling plan that picks up the 1000 bad cans pick up the defective caps also?

Finally, let us consider a reasonably common defect that can occur in making the flexible laminate: Some foreign matter breaks the smooth flow of molten polyethylene from the nozzle of the extrusion coater applying the heat-seal coating to the inside surface of the foil. This causes breaks in the coating for a length of 100 feet, these breaks crossing parts of all three widthwise repeats:

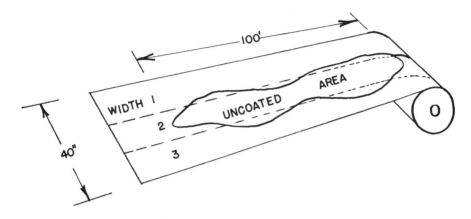

One hundred feet of laminate will contain 720 repeats, and of these, 240 defective repeats will appear in each of three slit rolls (or six rolls of a cutoff happens to fall within the 100-foot coating skip). Although the total defectives are only 0.072% of the million repeats, the three defective rolls will each have 24% defectives. As compared with the cans and the plastic closures, what sort of sampling program will pick out this defect?

To get at the answers to sampling the lots of cans, closures, and laminates, we must make a few unpleasant but necessary assumptions and decisions: First, we must accept the fact that there will be some defects in almost every lot of anything, and we must decide what percentage we can tolerate. Just as it seems to be a natural law that no two of anything are exactly alike, there seems to be another that says nothing is perfect. Fortunately, when we buy packaging materials, we are not looking for perfection—just usability—but when a process goes "out of control," as in the examples we are studying, some of the units produced become unusable. For purpose of discussion, let us for the moment set an arbitrary 0.1% as the allowable level of defectives. In technical terms, we are specifying an "AQL of 0.1%"; that is, the Acceptable Quality Level for the lot allows no more than 1/10th of a percent of defective units (1 defective per 1000 pieces).

The second unpleasant piece of news is that no sampling plan short of 100% inspection can assure us of finding the true level of defectives. In common sense terms, this means that you can't find all the bad pieces unless you inspect every one. Since 100%

QUALITY-CONTROL PRINCIPLES

inspection is far too expensive or destructive, sampling plans were developed as compromises between inspection cost and the *probability* of finding the percent of defectives. Again, in simple terms, the closer we come to 100% inspection, the greater the likelihood that we will approach the true level of defectives; and the smaller the sample, the greater the risk of missing the defectives. Sampling plans are numerous, and the several techniques of single, double, multiple, stratified, and sequential sampling are all intended to help the quality inspector in his game of finding the defects under the many possible different circumstances.

The many plans for sampling to find defectives are best presented in Military Standard 105D, published April 29, 1963 by the U.S. Department of Defense. A look at a table from this document will be enlightening. This one (Table 5) covers single sampling for normal inspection—a simple procedure for ordinary situations. Note that in our examples, a lot of 1 million calls for 1250

Table 5 Excerpt from MIL-STD-105D, Table II-A: Normal Inspection, Single Sampling

Lot size	Sample size	Acceptable Quality Levels (AQL)											
		0.1		0.65		1.0		2.5		4.0		6.5	
		Ac	Re	Ac	Re	Ac	Re	Ac	Re	Ac	Re	Ac	Re
501–1200	80	Use plan below		1	2	2	3	5	6	7	8	10	11
1201–3200	125	↓		2	3	3	4	7	8	10	11	14	15
3201–10,000	200	0	1	3	4	5	6	10	11	14	15	21	22
10,001–35,000	315	1	2	5	6	7	8	14	15	21	22	↑	
35,001–150,000	500	1	2	7	8	10	11	21	22	↑			
150,001–500,000	800	2	3	10	11	14	15	↑					
Over 500,000	1250	3	4	14	15	21	22	Use plan above arrow					

AQL means "maximum percentage of defectives allowable in the lot." Ac means "lot is acceptable if the defective specimens found in the sample do not exceed the indicated number." Re means "lot should be rejected if the defectives found in the sample equal or exceed the indicated number."

samples to be taken, and since the AQL we have decided to use is 0.1, we see that we can accept the lot if we find three or less defectives in the sample, and we should reject it if we find four or more.

Now we are in a position to summarize the situation with respect to the three lots in question and to analyze the sampling process. The following table lists what results in defectives from the conditions postulated:

Item	Metal cans	Plastic closures	Laminate
Total order	1 million	1 million	1 million repeats
Defectives	1000	1440	720 repeats
% Defectives	0.1%	0.144%	0.072%
Lot makeup	1000 pallets, 1000 cans each	2000 cases, 500 caps each	1000 rolls, 1000 repeats each
Distribution of defectives	All in 2 pallets	1/12 of 36 cases	All in centers of 3 rolls or ends of 6 rolls

We are able to state immediately, in answer to an earlier question, that the pattern of defectives is not the same in all three lots, because of differences in the natures of the manufacturing processes.

As for the sampling procedure and its ability to find the defectives, it can be seen that 1250 specimens in a lot of 1 million requires a sampling of every 800th piece. Translated to the materials as listed in the table above, the following would be taken:

Materials	Sampling procedure	Defectives picked, probability
Cans	1 or 2 per pallet	1 for sure, 30% probability of getting a second
Closures	1 each box or second box	1/12 chance of taking a defective in each of 23 specimens
Laminate	Outside end of each roll (only 1000 samples rather than 1250)	3 for sure if defect was at cutoff; none if at roll centers; overall, 25% probability of finding 3 defective rolls

QUALITY-CONTROL PRINCIPLES 69

Thus, the sampling plan gives assurance of picking up only one defective can and no certainty of finding any defectives in the other lots. This matter is worth more detailed examination to learn why, and more important, what to do to improve the probability of picking up a better measure of the percent of defectives. To look at each lot separately:

A. Cans

If the pallets are numbered and the quality technician does his sampling rigorously, he will not pick his specimens only from the top tier or the outside layers of each pallet—he will estimate the position of every 800th can after the previous specimen and take it from the middle of the pallet if necessary. If he starts with Can #1 in the total lot, the 627th specimen will be in the bottom of the 502nd pallet and the 628th in the top half of the 503rd. Thus, if he samples *rigorously* according to the plan, he will surely find *two* defectives. This being less than the maximum of three allowable, he will pass the lot as acceptable. We as readers are in the privileged position of knowing that the lot does indeed contain 0.1% defectives, no more than the maximum allowable, and does merit acceptance. But the sampling does not indicate any such information. It says only that since less than four defectives were found, the lot *probably* contains no more than 0.1% defectives, but it might be as low as 0.040%. How good is this probability? Oddly enough, that question cannot be answered except by experience with many such lots of cans. MIL Std 105D contains a table which shows Operating Characteristic Curves for single inspection, which we are discussing here. It indicates that for lots of 1 million, containing 0.1% defectives, with an AQL of 0.1%, the above sampling of 1250 specimens will result in acceptance of the lot in 95 cases out of 100, provided, of course, that the acceptance is on finding three or less defectives in the sample.

Practical considerations in sampling must now be mentioned. The technician is assigned to take 1250 sample cans from 1000 pallets, which should be made up of every 800th can from the start to the end of the can-making run. Think what this demands! The pallets should be numbered in order of their assembly, they should be standing on the floor in the same order, they should all be accessible to the technician, he should have the time and ability to dig inside the pallets to get his specimens, and he should number each so it can be traced back to its pallet and position. Contrary to this Utopian q.c. situation are the contrary pressure to minimize storage space by stacking pallets two or three high, in blocks with no space between, to minimize time in sampling by limiting sampling to available pallet surfaces, without numbering pallets or cans. Add to this the fact that the lighting in most warehouses is not ideal for finding pallet numbers if they exist, and the sample becomes random rather than representative. Had the two pallets with the 1000 defectives been side by side and the samples taken from them numbered, the technician would have found two leakers from adjacent pallets in the production sequence, which would be a tipoff to sample cans between those two, and the true picture on defectives would be revealed in the second sampling. But in a random sample he might get none or several from the two pallets of defectives. If the latter, the whole lot may be rejected, although it meets the specified AQL, requiring supplier/customer negotiation and resampling or sorting to reliably establish the actual state of affairs.

The moral of this story, if there is one, is to do the sampling while the manufacturing sequence is in process, rather than trying to untangle mixed pallets later. It is easy to take every 800th can off the can-making line as it runs and number it. Then, when the two defectives are found one after the other, the corresponding pallets can be found hopefully faster. The overall picture is that it is better to do this q.c. testing in the supplier's plant; when it has reached the customer, it is too late for quality *control*—the lot has been all made—and all that can be done is a quality *audit*. Control implies examination of quality in time to take action when necessary.

B. Closures

Since the plastic caps came from a 12-cavity mould, there is no 1st or 500th or 100,000th piece, unlike the can study, above. Once dropped into cases of 500 in increments of 12, they lose immediate identification of moulding time. Each box, however, will represent a molding interval of 7 minutes and will have in it about 504 pieces made up of 42 "shots" of 12 caps each. Thus, there will be in each box 42 pieces from Cavity #1, Cavity #2, and

QUALITY-CONTROL PRINCIPLES

so on up to Cavity #12. The sampling technician has no choice but to sample a single cap from each of 1250 cases to make up his total sample. If he is conscientious, he will not sample randomly, but will take 1/12 of his total sample from each of the 12 cavities, so that the final composition will have 104 pieces from each cavity, with a couple of extras. We noted above that all the defectives will be packed into 36 of the 2000 cases. To get 1250 specimens from the 2000 cases the technician will sample 1250 cases, or 63% of them. At this rate, he will take 1 cap each from 23 of the 36 cases with defectives. This should be two caps from all cavities but one, assuring him of finding one defective and 11/12 probability of a second. If he samples randomly without regard for cavity numbers, he has no assurance of picking up even one defective; the probability of doing so is 1/12. This is an application of the statistical problem of a box of mixed black and white marbles, in this instance 11/12 of them white, and 1/12 black. If one were to pick 23 marbles blindly from a large lot of this mixture, the probability of taking a black one in any single selection is always 1/12. There can be no sure conclusion that one or two or more will occur in the selection of 23.

Now, to compare this case with that of the cans, we can observe that while rigorous adherence to the sampling plan assures two defectives being found among the cans, it will assure only one in the closures. Furthermore, we, the readers, know that the closures lot contains 44% *more* defectives than the cans lot and does *not* meet the specified AQL level of 0.1% maximum defectives. Why is the sampling plan less reliable in locating these defectives, even when rigorously applied? The reason is that while the cans represent one lot, the closures lot is composed of 12 sublots mixed together. The million cans all came off one can line, and their condition is the result of the performance of that line only. On the other hand, the million caps are an assembly of 12 lots of 83,333 each (each sublot being the production that came out of one cavity). Referring now to the sampling table, we see that to inspect for an AQL of 0.1% in lots of that size, it would be necessary to take 500 samples from each sublot, accept on one defective, and reject on two. This means that the inspector must sample caps from Cavity #1 from 500 boxes, #2 from 500 boxes, etc., for a total of 12 × 500 or 6000 samples. This means taking one cap from each cavity from each fourth box. If he follows this plan, he will find nine defectives in the 36 boxes that contain them, and the sublot will be definitely both identified and rejected.

The original sampling plan of 1250 did not recognize the true composition of the whole lot, and what it did in fact was to sample about 100 pieces from each of the 12 cavities, only 1/5 the sample size required to measure an 0.1 AQL in a lot of about 83,000 pieces. The moral of this story is to be able to define a lot properly for purposes of inspection.

C. Laminate

In this situation we know that the actual defect level is only half that of the closures and well below the allowable 0.1%. Practical considerations make it almost impossible to rigorously sample the whole lot according to the plan, however. There are 1000 rolls of material, and the sampling plan calls for 1250 specimens. The only part of a roll available for sampling without destruction is the outer end. Although strictly speaking only one printed repeat need be taken for inspection, in practice a foot or more of laminate is usually cut off.

Although this doubles or triples the number of repeats and more than satisifes the sampling plan, the extra repeats do not really add more information. Since no one will agree to the cost of cutting into a roll just to get an interior specimen, we shall have to be satisfied with the outside ends.

The defect, it will be recalled, is 100 feet long and affects three rolls. The rolls are all a bit over 400 feet long, so the probability is 25% that the defect will occur at a cutoff; if it does, it will show on the outside ends of three rolls and it will also exist at the core ends of three other rolls. If not a cutoff, the defect will not show at all on a sampling of outside ends. Thus, in summary, the sampling plan cannot be rigorously carried out; a sampling of the outside ends from all 1000 rolls was a 75% probability of showing no defects, and a 25% chance of showing three defects, with nothing in between. The maximum of three defects is the greatest allowable to accept the lot, but with some doubt that an additional 250 samples to make the required 1250 might add at least one more defect. This is a situation about which the inspector cannot act

QUALITY-CONTROL PRINCIPLES 73

with assurance, since the practical limitations put on the sampling simply prevent it. Thus, it is possible for a lot of material to look questionable because of inadequate sampling as well as the more common opposite, where it looks better than it should because of too few samples.

This brief discussion of sampling packaging materials for inspection is by no means comprehensive; lifetime careers have been devoted with great value in results to the economics of mass production. The facts that are highlighted are these:

1. Generalized rational sampling plans are available for application to the inspection of packaging materials. They are published and widely used and serve as good guides to the part of quality control that depends on evaluating representative material.
2. Practical considerations impose themselves on almost every sampling situation which require modification by good judgment of the theoretical sampling plan that would apply. Much of this judgment can be used on knowledge of the processes and how they affect the materials produced, as shown by examples above.

IV. QUALITY STANDARDS

The time has come to define *quality*. When used as a single term, the word is almost meaningless, connoting no more than the general idea of "the way the writer of the spec intended that it should be." Like *love*, *quality* communicates a favorable image with an emotional appeal, but needs some objective descriptions to be usable as a basis for constructive action. When a perfect package is made, it can be said to possess certain *favorable qualities* and to be free of certain *unfavorable qualities*; such as, it opens easily, it recloses tightly, the graphics are in register and free from scuffs and scratches, it is not a leaker, the dimensions are correct, etc. In this sense, quality connotes conformance to spec and freedom from the effects of damage or abuse. The next step in refinement of the definition is to state the specified factors that make up quality and the effects of damage that must be avoided. *Quality* then becomes the sum of qualities present which are desired and the absence of qualities not desired.

But this is not yet enough to let us establish a basis for action on quality *control*. Quality standards must be more than a list of wanted and unwanted qualities, because few packages are perfect, and almost every package will have one or more of a couple of dozen possible defects. Yet all cannot be rejected, and here it is necessary to state the obvious, namely, that some defects are more serious than others. The recognition of this fact opens the door to a classification of defects and the assignment of different weights or scores in assessing the acceptability of a package under inspection.

It is customary to set up three classes of defects, defined as follows:

Class A. Critical defects: those which prevent a package from performing its intended function of protecting its contents, or from compliance with applicable regulations, or which cause safety hazards in handling, use, or disposal.

Class B. Major defects: those which cause borderline functionality in a package or reduced identity by virtue of graphic defects.

Class C. Minor defects: those which impair the appearance but not the function of a package.

Examples of classification for several common package defects can be shown:

Class A	Class B	Class C
Cracked or chipped glass bottle	Bottle below minimum weight	"Stones" embedded in glass walls
Fracture in can end or body	Missing color in can lithography	Small dents and scratches on can
Missing liner in closure	Out-of-round	Cap color off standard
Puncture in pouch	Heat seal too narrow	Printing out of register
Unglued case flap	Case out of square	Washboard appearance on corrugate

QUALITY-CONTROL PRINCIPLES

One more step will put us into position to act on the overall quality level of a lot of packaging materials: we must decide what kind of defects and how many of them make a single package or part unacceptable and then how many unacceptable units make a lot unacceptable. Obviously, one Class A defect makes a unit unacceptable, but what about the others?

Experience has developed an arbitrary scoring system for Class B and Class C defects which can be adjusted to the specific situation and to the needs of the customer. For instance, it can be agreed between supplier and buyer that two Class B defects or five Class C defects merit the rejection of a unit. This system is equivalent to charging defects as demerits, wherein a unit may be failed for having a total of 10 demerits; one Class A defects counts 10, one Class B defect counts 5, and one Class C defect counts 2. An inspector can now operate with reasonable objectivity in the examination of packages or components, in that he has a list of preidentified defects, each of which is known to be critical, major, or minor, and each of which has a definite value in scoring. He can prepare a scorecard on which the inspection results of each lot of materials can be recorded. Samples of such scorecards are shown in the succeeding chapters.

To conclude this discussion of standards, it should be noted that there are two kinds of qualities to be inspected: variables and attributes. *Attributes* are "yes-or-no" kinds of qualities such as were discussed in the sampling of lots of 1 million cans, closures, and flexible laminate repeats. The cans either leaked or they did not; the closures were completely formed or not; and the laminate either had a heat-seal coating or not. Variables are those qualities which can be measured to determine whether they are within or outside of an acceptable range. This *acceptable range* constitutes a quality standard; for example, the acceptable weight of a 1-quart glass bottle may lie in the range of 12.5 to 13.5 ounces. Bottles weighing below the minimum may be too fragile, and those above the maximum would incur freight penalties in distribution.

The purpose in making the distinction between variables and attributes as quality factors is not academic. We have illustrated the use of MIL-STD-105D to select sampling plans for inspection of lots of materials. Those plans are designed particularly for inspection of attributes. The inspection of variables may be conducted on sampling plans which have their basis in the theory of mathematical statistics. It is not our purpose here to review that theory, which is covered very capably in dozens of texts on statistics; the important fact is that any sampling plan for attributes will have greater reliability in representing the average and range of variables in a given lot. In theory, if packaging materials could be inspected for variables only, and no attributes, then smaller samples could be taken, but this is never the case, so it is simplest to use the "attributes sample" for both kinds of inspection.

Quality standards for variables are naturally numerical, since they represent measurable quantities. Standards for attributes may be verbal (no punctures, for instance) or visual (color standards, roughness limits). The former are straightforward, objective, and unequivocal, while the latter, being subjective criteria, may become quite troublesome. Limit samples offer the best available guidelines for making decisions on the acceptability of visual defects. Examples where limit samples can apply as standards include mottling in printed colors, streakiness in glass containers, deviations from color standards, surface texture of plastic parts, brightness of can-making plate.

V. ADJUSTMENT OF MANUFACTURING PROCESSES

It was noted previously that quality *control* can be undertaken only when sampling and inspection are done in time to correct errors in ongoing production. Such error correction may be the elimination of defective materials from the process or adjustment of the process itself to eliminate defect-causing conditions. An example might be the appearance of black specks in moulded thermoplastic tubs, found to be caused by the addition of dirty regrind. The defect can be corrected by leaving out the regrind, at a severe cost penalty. A better solution would be to change the whole scrap-handling and regrinding process to keep it free from dirt contamination. This may include collecting the scrap before it gets on the floor, storing it in covered containers, and grinding it into clean, covered containers.

When done as production is under way, quality control of packaging materials can be applied only by the supplier at his plant,

QUALITY-CONTROL PRINCIPLES

and his customer, the packer, can exercise quality control only over the packing process and his contained product. Any inspection made after a lot of anything is completed can only be a quality *audit*—it is too late to exercise control. The one option available at that point is to identify and cull out defective production. This truth makes for a very clear division of quality responsibilities between packaging-materials suppliers and their customers. The supplier is the only one who can be responsible for controlling the quality of packaging materials—the customer can only audit the quality of the materials. On the other hand, the customer is the only one who can control the packing process and thus control the quality of the finished pack.

The quality of a product is the result of incoming material and the process for converting the material into the product. This applies whether we are converting a raw material into packages or components, or whether we are packing products into packages. From what has been said on previous pages, it is clear that the estimation of materials quality is accomplished by inspection.

It is now time to talk about process quality, the other determining factor in product quality. Materials are *things* which can be inspected at leisure, to some extent, but processes are fleeting *activities*, whose conditions can change momentarily. Early in this chapter we examined the effects on lots of cans, plastic closures, and flexible laminate of short-term failures in process control. What is the best way to go about controlling processes? There are three means available:

1. Operator judgment
2. Operator judgment, assisted by visual/audible instruments or signals
3. Automatic control, with sensing instruments that feed back to control systems

Any of the three procedures may be supplemented by process inspectors who *audit* the functioning of the operators, instruments, or controls.

Operator judgment is not in fashion today, for some good reasons and some not so valid. The best reasons for downgrading operator judgment are that the process depends on the skills of an individual who may take some time to train, may be out sick sometimes, or may leave the job. On the other hand, an operator's job that requires no skill cannot hold the interest of most individuals; boredom will lead to inattention, and inattention to poor quality. From a technological standpoint, volume production processes often run at too great a speed for an operator to observe conditions and exercise judgment. For instance, timing a canning line that runs at a thousand containers per minute cannot be done without the aid of instruments.

Skilled operators who carry the full responsibility for process control are craftsmen, but few industrial processes today are run by craftsmen. The use of instruments to communicate process conditions became common decades ago. Oven thermometers help the operator judge whether papers and printing inks are drying properly. Heat-seal bars have thermometers to show whether good seals can be expected. Other gauges show coating thicknesses or weights, linear line speed. Such instrumentation extends the ability of the operator to use his judgment and improves the chances that his judgments will be correct.

The most advanced and growing system, of course, is the application of automatic process controls which maintain specified conditions. In an instrumentation or automatic control it is essential to know that the variable condition in question does in fact relate to product quality. For example, when drying a solvent-base overprint lacquer applied to a flexible laminate, the oven temperature is not nearly so important to control as the cfm airflow through the oven. Naturally, a thermometer is easier to install and maintain than an anemometer, but if the information it communicates is unimportant to the control of the process, it is misleading.

VI. PRODUCTION AT A MAXIMUM PERCENT OF ACCEPTABLE UNITS

It is a basic principle of quality control, as discussed previously, to maintain production in-spec, rather than to determine what percentage of the production must be discarded or reworked. The communications channels between inspectors and production supervision must therefore be short and always open. To minimize reaction time, face-to-face communication, telephone contact, and data sheets are the best reporting media. Typed reports and interoffice mail are inevitably after the fact, but good for

QUALITY-CONTROL PRINCIPLES

record purposes and management posting. An accumulation of such reports can often be used to justify automatic control installations by pointing up a problem that caused monetary losses over a period of time.

VII. MEASUREMENT AGAINST PRODUCT SPECIFICATIONS

We have now come full circle back to the specs in our discussion of quality control. All that has been said in this chapter is intended to describe quality control principles that keep production of a package "as the writer of the spec intended." One distinction remains to be stated: the difference between specs and standards. Specs must precede standards; as detailed in the first two chapters, the specs describe the structure and function of a given package. The standards, however, are the more precise targets and limits of structural and functional features which are used as a basis for in-spec/out-of-spec decisions in quality inspection. Let us clarify this distinction with an example, such as a beverage can with a ring-pull opening device on the top end.

The spec would say something like "The top end shall be formed of 0.012 inch aluminum, according to Print No.—. The dimensions on the print are part of this spec."

The quality control standards, however, would say this: "The residual score on the opening device is 0.003 plus or minus 0.0005 inch. The average maximum force required to open any can shall be 4 pounds, with a range of 3 to 5 pounds. Opening failures due to the ring separating from the rivet before effecting removal of the pour opening shall not exceed one per hundred cans inspected." Perhaps a shorter and equally effective statement would be "Specs are the basis for quotation, and standards are the basis for inspection." In the sample above, it will be noted that the standards include both variables and an attribute. Another attribute would usually be included, requiring that "the opening device shall not rupture when case lots of filled cans are subjected to laboratory rough handling to simulate distribution stresses. See Test Method No.—."

The chapters that follow will describe in more detail the specifications and quality-control methods suitable for common package types.

5
Glass Container Specifications and Quality Control

The elements of specifications for a glass package include several items:

1. A description of the verbal and numerical criteria related to the structure, function, and marketing requirements for the glass itself.
2. A print of the glass, with nominal dimensions, weight, and capacity and tolerances for all three. A supplementary print detailing the finish may be necessary.
3. A description of the closure, also accompanied by a dimensioned print. A graphic "mechanical" is usually separate from the technical spec, since graphics changes and technical changes are seldom simultaneous.
4. A description and blueprint of the label. As with closures, the graphic art is customarily separate.
5. A description of the structure and performance requirements for the shipping container. Again, the graphic "miniature" is separate.

It is easiest to explain by illustration from this point. Let us study a typical glass package spec, a 1-quart mayonnaise jar, cap, label, and shipper. The many packers of this product do not use the same spec, of course, but the one described here could be considered representative and in several respects a composite.

Following the presentation of the specs for the components of the package will be a discussion of the more interesting points, and behind that a consideration of quality-control matters in both suppliers' and packers' plants.

GLASS CONTAINER SPECS AND QUALITY CONTROL

SPECIFICATION FOR: <u>1-Quart Mayonnaise Package</u>   NUMBER: <u>714750</u>
PROPERTY OF THE XYZ COMPANY   EFF. DATE: <u>October 1, 1982</u>

I. <u>SCOPE</u>: This specification covers construction and performance requirements for a glass package to protect and distribute retail mayonnaise. It is for use at all plants, and replaces 71474 dated August 1, 1975.

II. <u>CONSTRUCTION</u>

   A. Jars shall be made of flint glass. A bottom stacking feature is required.

   B. Attached Print No. 520* is a part of this specification, with its dimensions, weight, capacity, and their tolerances. Wall thickness and glass distribution in the container shall be adequate for the packing and shipping stresses to which it is exposed.† Finished goods may be shipped by rail or truck.

   C. Deviation of the finish from the flat shall not exceed 0.015 inch in any 60° of the circumference, nor shall it exceed 0.030 inch around the entire 360°. The lip of the finish shall have a flat surface for sealing, not less than 0.030 wide, and preferably 0.050 inch.

   D. Out-of-round at the finish shall not exceed 0.030 inch. The label area shall be symmetrical so as to permit complete adhesion of spot labels. Jars shall stand squarely on their stacking beads. Leaners shall not exceed 1/8 inch on a side.

   E. Jars shall be free as commercially possible from defects such as bubbles, checks, chips, cracks, stones, stresses, and thin spots.

   F. Glass shall be free from contamination with any foreign matter that cannot easily be removed by washing before filling.

   G. Jars are to be treated on the outside with low-friction coatings. Formulas shall comply with applicable legal regulations and must not offer resistance to label adhesion. After approval by XYZ Company for product compatibility, no coating formulation shall be changed without prior notice, evaluation, and reapproval.

---

*See Figures 4(A) and 4(B).
†If the bottom panel formed by the baffle in the parison mould is swung outside the stacking bead, the possibility of thin sidewalls is indicated. See Sec. V, DEFECTS.

III. **PERFORMANCE**: It is intended that defects associated with breakage in packing operations shall be kept to the lowest possible commercial limits. Should the rate of breakage on a packaging line exceed 7 jars per 100,000 of any lot, due to identifiable Class A defects, the remainder of the lot shall be set aside for investigation, and another lot of glass shall be substituted.

Other conditions that make it impossible for XYZ Company to use delivered glass shall be cause for rejection, such as:

A. Jars wrong side up in shipper.

B. Wrong copy or illegible printing on shippers.

C. Bottom case flaps unglued.

D. Damaged cases which are unsatisfactory for outbound shipment.

E. Broken glass—any case found to contain any broken glass shall be rejected.

Complains on glass quality and/or performance found by XYZ personnel shall be communicated through channels within 24 hours to the Plant Buyer. Samples of allegedly defective glass shall be retained for inspection by the supplier.

IV. **INSPECTION**

A. For purposes of this specification, a lot shall be defined as a unit of delivery, most commonly a truckload of 15,000 to 30,000 jars.

B. The supplier shall certify that each lot has been inspected before delivery by sampling in accordance with MIL-STD-105D, and meets the following criteria:

Class of defect	Inspection level	AQL
A	Normal	0.65
B	S-2	1.0
C	S-2	6.5

Inspection for Class A defects shall follow Sample Size M in 105D. 315 specimens shall be selected from a lot, taking the specimens equally from all mold numbers represented, with acceptance up to

GLASS CONTAINER SPECS AND QUALITY CONTROL

five defects and rejection on six or more. The objective of the Class A specification is to achieve 95% assurance of less than 1% defectives in the jars received in any lot.

Specifications for Class B and C defects are intended to assure the operation of manufacturing controls on dimensions and appearance.

The supplier is considered responsible for conditions caused by his shipper which may make a lot received unacceptable.

C. Lots that fail the above inspection shall not be delivered by the supplier, who is expected to cull out defective material before shipment. If any lot fails the performance specification, it shall be inspected by XYZ quality inspectors in accordance with the plan specified in IV.B, above; should the lot be found within control, packing-line conditions shall be examined as possible causes. If the lot is found out of control, it shall be set aside for supplier reinspection, with notification to the Plant Buyer. Negotiation shall determine whether questionable lots are acceptable as a business decision, with provision for special attention in processing.

V. CLASSIFICATION OF DEFECTS
   A. Class A defects: those deviations from spec which prevent glass containers from performing their intended function of safely containing the products for which they were made, through a normal cycle of distribution and use, including:
      1. cracks and penetrating checks
      2. browouts and light sidewalls
      3. broken or chipped finished
      4. contamination with dirt, insects, or any other foreign matter except fiber dust which can be easily removed by washing prior to filling
   B. Class B defects: deviations from spec which render durability or packing-line fit questionable, including:
      1. glass weight below specified minimum
      2. functional dimensions outside of specified tolerance, such as height, diameter, capacity, and finish
      3. nonfunctional stacking bead
      4. leaners in excess of 1/8 inch on a side

C. <u>Class C defects</u>: attributes which adversely affect the appearance, but not the function, of the glass container, including:
   1. stones
   2. wavy appearance
   3. rough mould parting line
   4. uneven outer surface

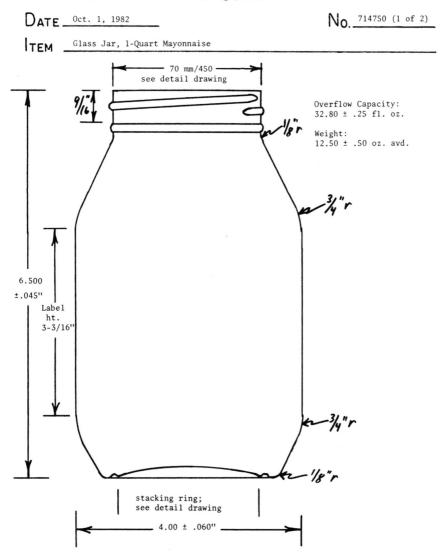

Figure 4(A)   Sample specification: Glass jar, 1-quart mayonnaise.

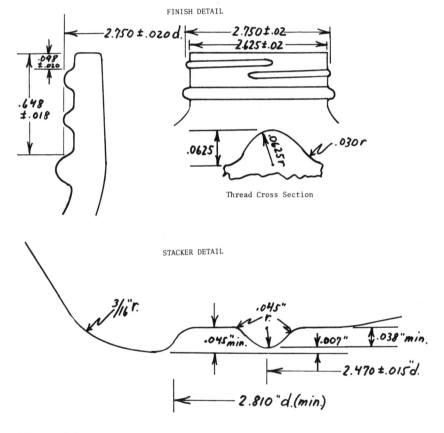

Figure 4(B) Sample specification: Glass jar, 1-quart mayonnaise; finish and stacker detail.

GLASS CONTAINER SPECS AND QUALITY CONTROL

SPECIFICATION FOR: Closure, 1-Quart Mayonnaise   **NUMBER: 714752**
PROPERTY OF THE XYZ COMPANY   **EFF. DATE:** October 1, 1982

I. SCOPE: This specification covers the construction and performance requirements for a continuous-thread metal cap used as the closure for 1-quart glass bottles of XYZ Mayonnaise. The corresponding glass specification is 71450, October 1, 1982.

II. CONSTRUCTION: The cap shall be made to fit GCMI Finish #450, 70-mm size. Dimensions and tolerances shall conform to Print No. 519,* attached, which is part of this specification. Other construction features shall be as follows:

   A. Steel used for making the cap shall be TFS/CCO†, 80 lb/base box, temper T-3.

   B. The exterior shall be lithographed according to separate graphic designs. A clear scuff-resistant varnish shall always be applied over the inks.

   C. Caps shall be lined with a food-grade pulpboard 0.040 inch thick, with a facing on the product side of a vinyl-coated paper impervious to the oils in mayonnaise, and containing no ingredients which will support bacterial or mould growth. On filled jars, caps shall hold nitrogen headspace flush.

   D. The cap shall have a stacking feature which matches that on the bottom of the jar.

   E. The interior side of the cap metal shall be coated with a food-grade varnish that is oil- and acid-resistant and gold in color.

   F. The liner disk shall have a minimum diameter of 2.700 inches, and the inside diameter of the liner retention bead in the metal cap shell shall be not over 2.680 inches nor under 2.660, so as to always hold the liner in the cap but not interfere with the "E" dimension of the glass finish.

III. PERFORMANCE: Caps shall be strong and rigid enough to resist distortion and damage in normal plant handling, including hoppering,

---

*See Figure 5.
†Tin-free steel, chrome/chromeoxide electrolytic coating.

sorting, rectifying, and application to glass at torque levels of 40 inch-pounds plus or minus 15. Frictional properties of the gold lining enamel and the liner facing shall be such that in normal distribution and handling, the removal torque is not lower than 5-inch-pounds less than the application torque.

IV. <u>DELIVERY</u>: Caps shall be delivered to XYZ Plants in corrugated boxes containing 1000 each. Boxes shall be end-loading style, sealed with one strip of reinforced tape at each end. Flaps shall not be glued. Boxes shall be stenciled:

1-Qt Mayonnaise Caps, 70 mm., Spec. No. 714752, 1000 pieces

Supplier's name   Lot No.   Date of manufacture

V. <u>INSPECTION</u>: Suppliers shall exercise sufficient quality control to assure compliance with this specification. Suppliers are held responsible for defects caused by their carriers in making deliveries to XYZ Plants.

Defects found by XYZ personnel shall be reported to the Plant Purchasing Agent through channels and via the Quality-Control Manager. If more than 1% Class A and Class B defects combined are suspected in any lot, the lot shall be set aside for reinspection and removal of the defectives by the suppliers. Class C defects shall not cause the lot to be discarded, but shall be brought to the attention of the supplier by the Plant Purchasing Agent.

VI. <u>CLASSIFICATION OF DEFECTS</u>

    A. <u>Class A</u>: those defects which prevent the closures from performing their functions of protecting and communication:
1. a crack in the metal anywhere on the cap
2. missing liner
3. warpage or out-of-round to the extent that cocking occurs in application to the glass
4. no threads or shallow threads that do not engage the glass threads
5. missing color or copy in lithography
6. dimensions outside of tolerance limits
7. incomplete or sharp edge on rolled bead

    B. <u>Class B</u>: defects which cause borderline functionality:
1. liners loose and partially hanging out
2. copy illegible in lithography or color off standard

GLASS CONTAINER SPECS AND QUALITY CONTROL

      3.   bulges or dents in the center panel of the cap

C. <u>Class C</u>: defects which impair appearance but not function:
1. blotchy lithography
2. scratches or scuff marks on outside surface of cap
3. rust spots from inadequate enamel coverage anywhere on the cap

PACKAGING MATERIAL SPECIFICATION

PROPERTY OF XYZ COMPANY

DATE Oct. 1, 1982   No. 714752

ITEM Metal Closure, 1-Quart Mayonnaise

1. CAP MAJOR DIMENSIONS

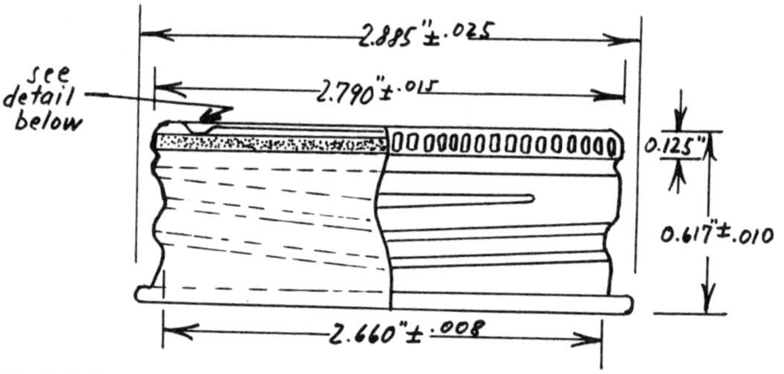

   - 2.885" ± .025
   - 2.790" ± .015
   - 2.660" ± .008
   - 0.125"
   - 0.617" ± .010
   - see detail below

2. STACKING PANEL DETAIL

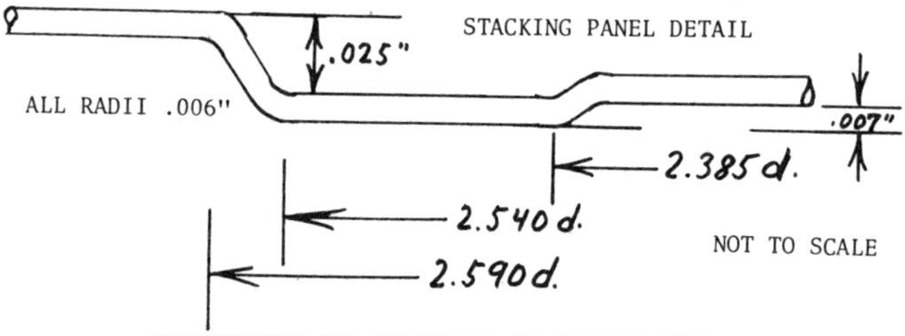

   ALL RADII .006"
   - .025"
   - .007"
   - 2.385 d.
   - 2.540 d.
   - 2.590 d.
   NOT TO SCALE

3. DESCRIPTION AND MATERIALS OF CONSTRUCTION

   Shell- 80T3, steel, TFS/CCO
   Liner- .040" pulpboard, food grade, with food grade vinyl greaseproof facing
   Inside lacquer- Gold #42, food grade
   Outside enamel- One or two colors on white per mech. art, scuff-res. varnish

Figure 5 Sample specification: Metal closure, 1-quart mayonnaise.

GLASS CONTAINER SPECS AND QUALITY CONTROL

SPECIFICATION FOR: Label, 1-Quart Mayonnaise   NUMBER: 714753
PROPERTY OF THE XYZ COMPANY   EFF. DATE: October 1, 1982

I. SCOPE: This specification covers the construction and performance requirements for paper labels to be applied on glass mayonnaise jars, 1-quart size. The corresponding glass specification is No. 714750, dated October 1, 1982.

II. CONSTRUCTION: Labels shall be made of 60-pound stock, litho-coated on one side. Dimensions shall be 3 inches high by 4 inches long, with rounded corners of 3/8-inch radius. One spot label shall be applied to each bottle. Tolerance ±1/32 inch. Printing may be done by any process that satisifes the need to match the original mechanical art.

Grain direction of the label paper shall be in the long dimension of the label. A water- and scuff-resistant lacquer shall be applied as an overprint finish.

III. PERFORMANCE: Labels shall be suitable for operation in any automatic spot-labeling machine which feeds from stacks. Stacked labels shall not stick together nor curl when stored at 40 to 65% humidity.

Tear strength by Elmendorf Tester shall be not less than 46 in the grain direction, nor less than 50 in the cross-grain direction.

Light resistance of the printing inks shall be not less than 10 Fadeometer hours.

IV. PACKING FOR DELIVERY: Labels shall be delivered in stacks of 500, the stacks to be firmly banded and even so that they may be code-dated by edge cutting. Each 10 stacks shall be cartoned, and four cartons packed for delivery in a corrugated RSC case.

V. INSPECTION: Suppliers are expected to exercise adequate quality control to assure compliance with this specification. The supplier is considered responsible to settle defects caused by his carrier in making deliveries.

XYZ personnel shall report defects found to the Plant Quality-Control Manager, who shall in turn report them to the Plant Purchasing Agent. Stacks of labels found with Class A defects, as defined below, should be rejected for credit to the supplier on the authority of the Plant Quality-Control Manager. Stacks found with Class B defects shall be set aside for further inspection and sorting of defectives by the supplier. Class C defects shall be reported to the supplier. The classification of defects is as follows:

A. Class A : defects which prevent the labels from functioning in the labeling equipment or from properly identifying the product:
   1. wrong copy or size
   2. missing color or colors
   3. labels in stack stuck together
   4. severe curl, such as not to apply in labeler
   5. tears or holes
   6. uneven stack, such as not to feed in labeler
   7. smeared or illegible copy

B. Class B : borderline conditions of appearance or functionality:
   1. rough-cut edges
   2. poor color registration or cut off-center with respect to printing
   3. scratches or below-standard scuff resistance
   4. colors off-standard or mottled

C. Class C : minor defects:
   1. loosely banded stacks
   2. stack count off by more than 3%
   3. improper or illegible identification on cases

PACKAGING MATERIAL SPECIFICATION

PROPERTY OF XYZ COMPANY

DATE Oct. 1, 1982   No. 714753

ITEM Paper Label, 1-Quart Mayonnaise

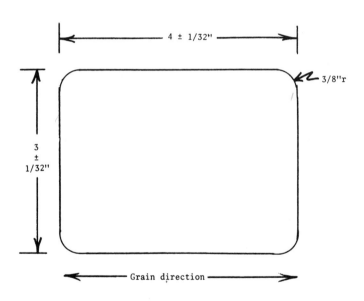

Label stock: 60-pound, litho coated one side

Figure 6 Sample specification: Paper label, 1-quart mayonnaise.

CHAPTER 5

SPECIFICATION FOR: Shipper, 12/1-Quart Mayonnaise **NUMBER: 714751**
PROPERTY OF THE XYZ COMPANY   **EFF. DATE: October 1, 1982**

I. <u>SCOPE</u>: This specification covers construction and performance requirements of a corrugated box used for receipt and distribution of 12 jars of 1-Quart XYZ Mayonnaise. The related bottle specification is <u>No 714750</u>, dated <u>October 1, 1982</u>.

II. <u>CONSTRUCTION</u>: The shipping case shall be RSC style, 4 X 3 X 1 arrangement. The outer shall be made of 200-test board, and the partitions shall be double-wall nontest, full-height, with shorts over longs for automatic uncasing.

Other construction features shall be:

A. Manufacturer's joint glued.

B. Case and partitions shall be made of C-flute corrugated, with flutes in the vertical direction.

C. Top and bottom flaps shall be treated with a nonskid coating to impart pallet stability.

D. Case inside dimensions shall be 16-5/8 X 12-7/16 X 6-3/4. Tolerances ± 1/32 inch.

III. <u>PERFORMANCE</u>: Cases shall be manufactured and delivered with consistent quality which permits processing on automatic uncasers, casers, and palletizers. Empty bottles shall be packed upside down for delivery in cases with bottom flaps sealed and top flaps folded shut but not glued.

IV. <u>INSPECTION</u>: The supplier is expected to exercise sufficient quality control over his cases to meet this specification and is also considered responsible for defects caused by his carrier in delivery to an XYZ plant.

Personnel in XYZ Receiving and Packing Departments shall report case defects to the Plant Quality-Control Department, who shall inspect complaints and if appropriate shall advise the Plant Purchasing Agent. Deliveries with Class A defects may be rejected; those with Class B defects shall be set aside for inspection and removal of defective units by supplier; and those with Class C defects may be packed and shipped, but notification shall be made to the supplier.

V. <u>CLASSIFICATION OF DEFECTS</u>: The classification of case defects is as follows:

GLASS CONTAINER SPECS AND QUALITY CONTROL

A. <u>Class A</u> : defects which prevent shippers from safely containing the primary packages through national distribution by common carriers, or which do not correctly and clearly identify their contents;
   1. loose manufacturer's joint
   2. dimensions outside of tolerance limits
   3. any break or cut through liner at a score
   4. wrong or illegible copy, or wrong color graphics
   5. loose partitions which fall out in uncasing equipment
   6. contamination with any foreign matter which stains or adheres to glass
   7. any tears, punctures, holes, or ragged flap edges with loosely adhering board particles
   8. bottom flaps unglued, or top flaps glued
   9. torn or bent partitions

B. <u>Class B</u>: defects which make function questionable or borderline:
   1. incomplete or partially illegible printing
   2. bends other than on a score line
   3. liner not completely glued to corrugated medium
   4. gap greater than 3/16 inch between matching edges of long flaps when box is set up
   5. absence of nonskid coating on top and bottom flaps
   6. moisture content under 5% or over 30%

C. <u>Class C</u>: defects which impair appearance, but not function:
   1. light or blotchy graphics
   2. uneven color or stains on outer surface, scratches, or scuff marks

The supplier is expected to exercise sufficient quality control so that no cases are delivered with Class A defects, not more than 1% with Class B defects, and not more than 5% with Class C defects.

PACKAGING MATERIAL SPECIFICATION

PROPERTY OF XYZ COMPANY

DATE   Oct. 1, 1982   No. 714751

ITEM   Shipper for 12/1-Quart Mayonnaise

Top — Inside dimensions, all ±1/32"

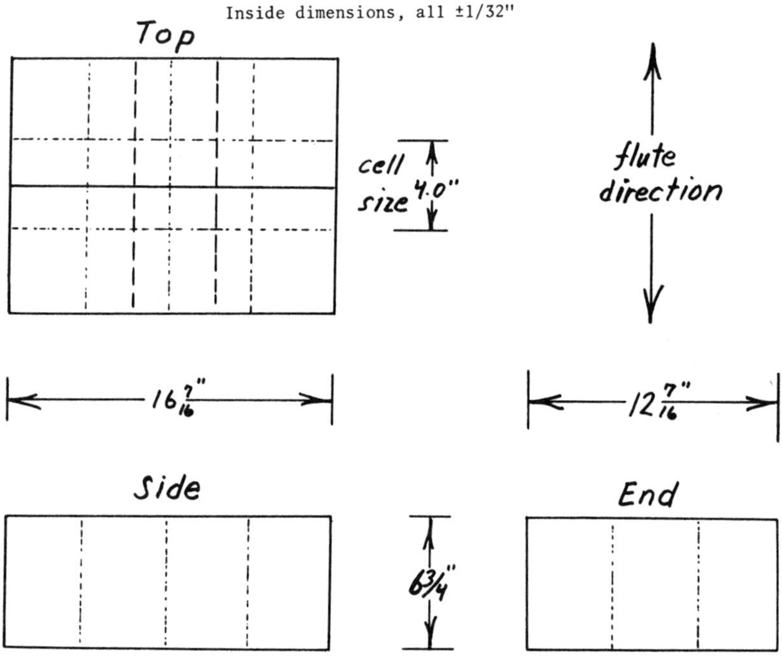

cell size 4.0"

flute direction

$16\frac{7}{16}$" $12\frac{7}{16}$"

Side $6\frac{3}{4}$"   *End*

Description: RSC style, 200-test natural kraft board, 42 x 42 liners. Non-test double-wall partitions, full height, vertical flutes. Printed one or two colors, dep. on mech. art.

Figure 7 Sample specification: Shipper for 12/1-quart mayonnaise.

GLASS CONTAINER SPECS AND QUALITY CONTROL

Now that we have looked at a typical spec format for a glass package, it will be interesting to comment on some of its important features, of which there are many.

1. Print #520 shows a notation at the top, "70 mm—#G-450," with arrows spanning the finish. This notation identifies a detailed spec for the finish, including the thread configuration, its dimensions, its inclination from the horizontal (helix angle), and tolerances for all quantities. A moment of reflection will bring to the mind of any buyer the desire that he should like to be able to purchase the mayonnaise bottle and cap from any of several suppliers and have them operate interchangeably on his packing lines and in the marketplace. This is in fact possible and is made so by the Glass Packaging Institute, a trade association of many glass and closure manufacturers, commonly known by the abbreviation GPI. Its members have agreed on standards for a series of glass finishes to prevent a chaotic proliferation of one-of-a-kind finishes which differ only slightly in diameter or thread angle or clearance between the lip and the top of the thread, etc.

2. The stacking feature at the bottom of the glass—a shallow ring—has dimensions that match those of the panel on the top surface of the metal cap. This interlock, which occurs when one filled jar is stacked on top of another in a retail display or supermarket shelf, contributes greatly to the stability of a stacked arrangement.

3. Each major component of the total package—glass, closure, label, and shipper—has in its own spec a combination of structural and functional requirements, as it should be. Furthermore, defects to be avoided are listed for each and classified as to their seriousness. Somewhere in each spec the intent is stated; for instance:

Glass: "... to protect and distribute retail mayonnaise."
Closure: "... caps shall hold nitrogen headspace flush ... shall be strong and rigid enough to resist distortion and damage...."
Labels: "... labels are intended to be clean, glossy, and convey quality."
Shippers: "... for receipt and distribution of 12 jars ... which permits processing on automatic uncasers, casers, and palletizers."

4. As noted earlier, the specs are written with the viewpoint of the packer, the buyer of the packaging materials. The classification of defects is based on his evaluation of their relative importance to his operations, costs, and market. The quality levels he can accept are communicated in the spec to the suppliers of the packaging components.

On the other hand, only the suppliers can control the quality levels of their production, and their q.c. efforts must be aimed at the customer's AQL's. This situation can obviously lead to problems if communications between buyer and seller are not open and continuous. But in its final analysis it is a situation no different from any other arrangement for buying by specification. If the packer were buying a stock bottle, he will generally assume that he buys on the supplier's spec, but when he buys to his own spec, both parties must be more flexible.

In this mayonnaise jar spec, there is nothing controversial about the items in II. CONSTRUCTION of 714750, since any glass manufacturer will agree to the statements and figures included. The statement in III. PERFORMANCE on wanting a breakage rate not to exceed seven jars per 100,000 on the packing line, however, is a very different matter.

Should any given lot of glass exceed that rate, it might equally well be caused by packing-line conditions as by faults in the glass; for this reason the buyer cannot put the statement into his spec as a *requirement*. There is nothing wrong, however, with stating his *intention* as a desirable objective, and in IV. INSPECTION, the procedure for handling such a situation is spelled out.

How did the packer arrive at a figure of seven per 100,000? This number could be approached from two directions; let us study both. First, there are production standards to be met. Let's say that the mayonnaise line runs at 240 jars per minute (20 cases), and that a running time of 6 hours per 8-hour shift is "standard." Expected output per shift is then 86,400 jars (7,200 cases). Each broken jar on the line may take 5 minutes of downtime to thoroughly clean up. At a rate of seven breaks per 100,000, six breaks could occur in a "standard" output of 86,400 jars, and would account for 30 minutes down-time. The six-hour running time in an 8-hour shift allows 1/2 hour for lunch, two 15-minute breaks, and 1 hour for all other causes. If six jar breaks

GLASS CONTAINER SPECS AND QUALITY CONTROL

take up half of that latter hour, only 1/2 hour is left for all other causes. Then, if additional breaks occur, they will cut into production at the rate of 100 cases (1200 jars) each. Thus, the figure of "7 jars per 100,000" represents all the down-time for breakage that can be tolerated in making "standard" production per shift.

The other way of looking at this figure comes from the AQL of 0.65 for Class A (breakage-related) defects. Superficially, the AQL says that there may be 650 jars per 100,000 delivered with such defects. Conditions on the packing line will not, of course, seek out each of the defects and produce a break, but if one defect of 100 were to result in an actual break, that would be seven breaks per 100,000.

Packing-line abuse varies from one line to another, of course. If jars never touch one another, or metal surfaces, nor undergo more than two gravities of acceleration in starting, stopping, shunting, lifting, or dropping, weak spots will be stressed only in the capper and the labeler. Breaks will be much lower than 1 per 100 defects. Should the line have many twists and turns, long surges, metal conveyors, and sudden starts and stops, breaks will occur much more frequently as a percentage of defects. On the whole, one break per 100 defects is probably not a bad guess for the average line.

Getting back to the mayonnaise jars, it will be noted in IV.C that the inspection plan is the deciding factor; the spec calls for the supplier and the packer to both participate in the examination.

I. QUALITY CONTROL OF GLASS PACKAGING MATERIALS

1. The main component in a glass package is, of course, the glass itself, so our discussion of quality control will start there. The first act in making glass containers is to mount a set of moulds into a glass-blowing machine. Next, the moulds must be heated and the glass "gob" weight adjusted to fall within the specified range. Finally, the machine is started, and glass is sampled from each mould, checked for all three classes of defects right at the machine, and returned to cullet until the process "settles down" into specification with respect to bottle attributes and variables.

The reasons for settling-down period are analogous to the situation of starting a car on a cold morning. It takes 20 to 30 minutes of operation before the moulds and related parts of the forming machine are thoroughly warmed, carbonized lubricant is burned off, and continuous draw from the glass feeder has stabilized gob temperature and viscosity. During the startup, all of the glass formed is discarded to cullet after inspection by the machine operator.

When the operator is satisfied that the process is in spec, he starts feeding the formed glass containers into the annealing lehr. He continues to inspect glass off the forming machine, from one mould after another, and should the pieces from any one mould start to run out of spec, he can stop the feed of glass to that mould until the problem is corrected.

The machine operator is responsible for culling bad glass and for controlling his equipment to produce a maximum output of in-spec glass throughout his shift. Beside each glass-forming machine is a bench with gauges and a spec chart against which he checks the glass coming off the forming machine. The bench also contains tools for adjusting the forming machine and spare mould parts in case of need for replacement.

Because the "settling-down" process at startup yields no salable containers, no more stops are made for shift changes, etc., until the total order for a given container is filled. Glass-making operates around the clock every day of the year, with several forming machines running off the same glass furnace so as to maintain dynamic equilibrium in the melt temperature. When one machine is down for mould change, the others are running; a furnace is shut down only once every couple of years to reline it with new firebrick.

Other than checking dimensions with gauges, the machine operator is trained to look for defects in container attributes, relate them to possible causes, and adjust his machine to correct them, or notify the proper person if the cause is outside his control.

Here are a few examples of conditions that a forming-machine operator can observe and correct:

Defect	Cause	Action
Rough parting line	Worn or chipped mould	Replace mould

GLASS CONTAINER SPECS AND QUALITY CONTROL

Defect	Cause	Action
Black specks on outside of bottle	Mould lubricant carbon residue	Heat mould to burn off residue
Stones	Disintegrating firebrick in furnace	Advise supervisor
Wavy inside surface of bottle	Parison too cold	Increase parison mould heat
Sagged finish	Blow mould too hot	Reduce blow-mould heat
Flat side on bottle	Blow mould not venting	Clear plugged vent or replace mould
Bottles underweight	Glass melt too cold or gob cutoff too fast	Adjust gob cutoff

Each forming machine is visited periodically by a quality inspector, who checks off-coming glass with the operator's gauges, and who also takes a specimen from each mould running. He returns it to a Control Laboratory, which is usually a room in a nearby part of the plant with glass-testing equipment. There, the samples picked up from the forming machines are checked more rigourously for weight and dimensions, but also for internal capacity, wall thickness distribution, leaners, out-of-round, and checks.

Should any variable or attribute show up in the Control Laboratory as out of control, the technician will notify the machine operator to take appropriate action. Since a forming department may be running a dozen or more machines at a time, each with a different bottle spec, the Control Lab staff must work fast to minimize reaction time. The slowest forming machines make 60 glass containers a minute, and fast ones can produce over 200 a minute. A delay of 1/2 hour in correcting a defect on a 100-per-minute forming job will let 3000 containers go by. Should 1 of 12 molds be running out of spec during that time, 250 defective containers will have been produced.

To speed up their inspection, the Control Lab staff uses an optical comparator, a device which projects an enlarged silhouette of a container onto a screen, which simultaneously projects a dimensioned transparency of the spec print. The technician can instantly see whether the container "shadowgraph" matches the print by adjusting one image to fall on the other, or if they fail to match, he can see where the container fails to match and by how much.

Wall thickness distribution is checked by cutting a cross-section of the container with a diamond-abrasive wheel.

After forming, the next factor important to glass quality is the annealing process. Controlled cooling in a long, tunnellike oven minimizes residual stresses inside the glass that would otherwise result from rapid cooling after forming. The quality of annealing is controlled by the oven conveyor speed and the temperatures in the several sections of the oven. The results can be checked qualitatively by visual inspection of the cooled glass containers with a polariscope. Looking through a bottle with polarized light, one will see bands or patches of strong color where the glass has unrelieved stresses, and almost no color if the container has been properly annealed. A numerical scale which refers to photographs of "limit samples" aids the inspector in making a decision on the adequacy of annealing in the specimens given him.

When the containers emerge from the cool end of the annealing lehr, they are either fed into a single-line conveyor for inspection before loading into trays or cases for warehousing, or inspected by the packers who take them off the lehr and load them directly. Whichever procedure is followed, the purpose is to cull defective glass; all inspection and packoff stations have hoppers for disposal of defectives.

Glass that has passed inspection and been packed is palletized—up to this point the glass has been a continous stream of containers; once packed it can be segregated into identified lots whose quality can be inspected according to a statistical plan. The Finishing Department, as the final inspection and packing area is usually named, invariably has another Control Lab which inspects finished ware for defects in both glass and shipper.

Since the XYZ Company has a spec for its 1-quart mayonnaise jars, in which quality levels and defects are clearly defined, the glass supplier can inspect ware accordingly. Figure 8 shows a possible form of lot inspection sheet which can be used by either seller or buyer or both to check the glass. The form relates to what has been discussed previously in the following ways:

GLASS CONTAINER SPECS AND QUALITY CONTROL

```
                    PLANT CONTROL LABORATORY
            PACKAGING MATERIAL LOT INSPECTION RECORD - GLASS

        Lot # 75        Date 2/2/83   From Crystal Glass Co.
        For 1-qt. mayonnaise              Spec # 714750
        Lot size 2000 cs. (24,000 jars)   Sample size 315

                            DEFECT COUNT

         Class A              Class B              Class C
   1 defect=1 defective   2 defects=1 defective   5 defects=1 defective

         AQL 0.65              AQL 1.0              AQL 6.5

   Cracks, checks //    Low weight      0    Stones ////
                  2                           ///             8

   Chipped finish  0    Off dim.        0    Waves //// //// ////
                                              //// ////
                                              //// //// ////    40

   Blowouts        0    No stack        0    Rough p.l. //// ////
                                              //// //             17

   Contaminated    1    Leaners ////    5    Rough surface   0


   Total defects   3           5              65
                               2              5
   Defectives      3          2.5             13
   Limit           5           7              21

   Action:    Accept  ✓   Reject____   Re-inspect____  Sort____
   Line performance (as reported by Production)  OK
   Comments Both checks from Mould #14
   Rough parting lines from Moulds 7 & 12          by J.J. Smith
```

Figure 8 Sample glass inspection form and tally of a typical lot.

1. A truckload quantity is given a lot number, with a date (shipped or received), XYZ Company spec number, and a sample size from Table 5 (Chapter 4).
2. AQL levels and point values given Class A, B, and C defects are used as described earlier in this chapter to "weight" individual defects and calculate an equivalent in total defectives.
3. The limiting number of acceptable defectives is taken from Table 5 and shown as a basis for deciding whether the lot is acceptable.

To use the form, the inspector fills in a hashmark for each defect found and then adds up a total for each type of defect in his sample of 315 pieces. His comments can be valuable feedback if communicated to the Forming Department.

See also Figure 16 and related discussion in Chapter 7.

2. Metal closure quality control is somewhat simpler than that for glass, because there is no "startup period" problem. The supplier must inspect the steel he receives for gauge, temper, and sheet or coil size. His first act is to cut proper-size sheets and coat one side with a lacquer or enamel which will be the inside coating after the caps are formed. The sheets are then lithographed with the customer's design, and the first quality inspection is that for color match and registration. Each sheet may carry several dozen circular repeats of the cap design, depending on cap size and litho equipment limitations.

A gang die on a punch press then stamps out circles from the printed steel sheet, and the circles are drawn into cup shape. At the next stations in a cap-forming machine, the drawn caps are threaded, knurled, beaded, and lined with coated-board circles stamped out on a secondary die.

What kinds of defects can occur in this process? To name a few:

1. Cap metal can crack in drawing if the die is rough or if the steel sheet contains a flaw.
2. Threads and liner retention bead can be too shallow if dies become worn.
3. Liners can be missing or only partially inserted behind the retention bead if liner roll stock runs out and a new roll is not properly fed in.

GLASS CONTAINER SPECS AND QUALITY CONTROL

    4. A sharp edge can form on the rolled bead if the drawing and beading tools get out of adjustment.

These are Class A defects which, as in the case of glass containers, can be listed with the other two classes of defects on an inspection form, such as in Figure 9.

The sampling plan for metal closures should recognize that a very small number of dies is required to stamp and form them, as compared to glass containers, which are formed in a dozen or more moulds. The advantage is that caps have a much lower probability of tool variables, but a corresponding disadvantage is that if one die fails, half or more of the total output will be defective until the machine is stopped. In glass forming, if one mould is defective, only 1/12 or so of the jars produced will be defective.

Another factor useful to the control of cap forming is that thread and heading rolls wear smoothly, and a chart of part dimensions versus time will show prior warning that a limit is being approached and the tools need replacement. Other than out-of-spec dimensions, including shallow threads, metal cap defects are likely to be random rather than systematic.

    3. Label quality control is mainly concerned with appearance factors, ease of handling, and operability on labeling equipment. With respect to the defects noted in Spec #714753, some comments on control may be helpful on items that are not obvious. The following two examples are taken from the list of Class A defects:

    3. "Labels in stack stuck together." This would be caused by insufficient cure in the overprint varnish, due in turn to low temperature or air flow or insufficient time in the curing oven. Process inspection is of course the basis for maintaining control, and a tack test on the finished labels is a suitable audit.

    4. "Severe curl." Lack of moisture control in the label paper is the cause. If the labels curl toward the unprinted side, the paper is too dry. Ink and varnish stabilize the dimensions of the printed side, so that expansion and contraction of the paper in response to moisture changes must cause curl toward and away from the printed side respectively.

PLANT CONTROL LABORATORY

PACKAGING MATERIAL LOT INSPECTION RECORD - METAL CLOSURES

Lot # **88**   Date **2/8/83**   From **Plated Steel Corp.**
For **1-quart Mayonnaise**   Spec # **714752**
Lot size **450 cases (450,000 caps)**   Sample size **800**
Case identification check **Stencilling ok**
Count check, caps per case: 1. **998**   2. **1005**   3. **997**

DEFECT COUNT

Class A 1 defect 1 defective	Class B 1 defect 1 defective	Class C 2 defects 1 defective
AQL 1.0, Class A and Class B		AQL 6.5
Cracks **0**	Loose liners **卌 5**	Blotchy litho **0**
Missing liners /* **1**	Illegible **0**	Scratches, scuffs **||| 卌 卌 卌 卌 23**
Warped **||| 3**	Off color **0**	Rust **0**
Shallow threads **0**	Bulges, dents **|| 2**	
Missing color **0**		
Off dim. **0**		
Sharp edge **0**		
Total defects **4**	**7**	**23** **2**
Defectives **4**	**7**	**11.5**
Limit **14, Class A + B combined**		**21**

Action: Accept **✓**   Reject ____   Re-inspect ____   Sort ____
Line performance (as reported by Production) **a few cocked caps**
Comments ***taped joint, roll change***   by **R. X. Jones**

Figure 9   Sample cap inspection form and typical lot tally.

GLASS CONTAINER SPECS AND QUALITY CONTROL

It is not necessary to go into detail on an inspection form, but a short sampling discussion may be interesting. Let it be assumed that a lot is 1 million labels. The lot will be composed of 50 boxes, each containing four cartons (200 cartons) and the cartons holding 10 stacks each (2000 stacks).

If we are concerned solely with labels, Table 5 says that a sample of 1250 should be taken and inspected as individual specimens. Several quality criteria in Spec #714753, however, lead to a listing of stack defects. For the inspection of stack defects, therefore, Table 5 says that from a lot of 2000, a sample of 125 stacks should be taken. Overall, the most effective plan would be to sample the 125 stacks first, inspect them for stack defects, then take 10 labels from each stack to make up the label sample of 1250, and inspect them singly.

An important point to be made here is that the lot-sampling plan is not necessarily unique; there may be two or more options, depending on the manner in which the units are subgrouped to make up the total lot.

The inspection and quality control of shipping containers will be covered in a later chapter.

6

Metal Container Specifications and Quality Control

In contrast to glass packages, metal cans have fewer components and less variety in shape, lots of a given spec are made on a single former, rather than a set of moulds, and dimensional tolerances are considerably smaller. On the other hand, while most container glass is composed of a single basic raw-material formula of silica, sodium carbonate, and calcium oxide, steel for can-making offers a choice of over 80 combinations of weights, tempers, and coatings. Aluminum, in addition, makes available a couple of dozen material specs which vary in alloy, form, gauge, and hardness.

This circumstance gives immediate notice that the sources of variability in cans are very different from those in glass packages, and that the quality-control program must therefore be approached very differently.

First let us look at the spec for a typical steel can, commonly called "No. 2," used for packing fruits and vegetables, with dimensions 307 × 409 (3 7/16 inch diameter, 4 9/16 inches high). Its capacity is about 20 fluid ounces, and it will contain 20 avd. ounces of pineapple, beans, or soup, for instance.

As with most cans, the spec shown is not the only spec for a No. 2 can. Several variations are commercially produced of the following structural features:

1. Weight and temper of body plate
2. Body beads, including none
3. Tin coating weight, including none, if tin-free steel is used
4. Enamel lining system, including seam stripe
5. End profile, plate weight, and temper

The variations in (1), (2), (3), and (5) arise from differences in can fabrication equipment at the plants of the several manufacturers of the No. 2 can, while the others are specific to the corrosive properties and shelf-life requirements of the products contained.

METAL CONTAINER SPECS AND QUALITY CONTROL

SPECIFICATION FOR: 20-Ounce Bean Soup Can   NUMBER: 520810
PROPERTY OF THE XYZ COMPANY   EFF. DATE: November 1, 1984

I. SCOPE: This specification describes structure and performance requirements for a tinplate can to protect and distribute retail brands of 20-ounce XYZ Bean Soups. It is for use at Plants 3 and 4, and is a new spec, not replacing any prior.

II. CONTRUCTION
   A. The can shall conform to the commercial specs for a No. 2 can and shall contain 20 net ounces of product. Nominal dimensions are 307 X 409.
   B. Body shall be made of 0.25 ETP (soldered), or CCO/TFS (welded), and the plate weight may be 90TU or 90T5 unbeaded, or 80TU or 80T5 with six beads. See Print No. 454,* which is part of this specification.
   C. Ends shall be made of electrolytic chrome/chrome oxide (CCO) tin-free steel (TFS), and the plate weight shall be 90T3. See Print No. 455 for end profile.†
   D. Epoxy enamel lining shall be applied to both inside and outside of ends and body, with seam stripe. The specific enamel formulation shall be left to the responsibility of the can supplier, subject to the following conditions:
      1. Shelf life shall be not less than 24 months at 70°F, as determined by XYZ Company product evaluation methods, which will be described to potential suppliers in invitations to bid.
      2. Supplier must have on file with XYZ Company an affidavit that the enamel and its ingredients comply with all applicable pure food laws and regulations.
      3. Lining compounds shall not cause undesirable flavor effects in the soup products as a result of normal processing and storage, as judged by XYZ Company Research Department evaluation methods.
      4. Having received approval for a lining formulation, a supplier shall make no change whatever in its composition or applica-

---

*See Figure 10.
†See Figure 11.

cation, without submitting new samples with the proposed change to XYZ Company for evaluation, and without having received written approval for the proposed change based on such evalution.

III. DELIVERY

    A. Cans shall be delivered with one end attached and the other end flanged.

    B. Delivery of the cans shall be palletized, using 48 X 40 inch of any construction with four-way entry for fork-lift handling. Cans shall be in 10 layers, 14 X 12 cans per layer, with solid chipboard separators between layers, and also under the bottom layer and over the top layer. Each pallet shall be securely shrouded for integrity and cleanliness of the load, with either corrugated board or shrink film. Cans shall be stacked on the pallet with open end up.

    C. Loose ends shall be delivered in stacks of 500, in corrugated boxes. Each stack shall be in a loose-fitting paperboard sleeve, and there shall be four stacks per box. Boxes of ends shall be delivered on 48 X 40 pallets, one box high, with stacks upright, and boxes strapped together.

    D. Pallets on which cans and ends are supplied may be one-way or returnable, at the option and mutual agreement of each XYZ Plant with its local can supplier plant.

IV. INSPECTION: Suppliers shall maintain adequate quality-control effort to assure compliance with this specification. Since can quality as received is important to the XYZ Company canning operations and finished product quality, the supplier shall be considered responsible for damage to empty cans caused by his carrier in making deliveries. XYZ Plant personnel shall report defects observed to the Plant Quality-Control Manager. He may at his discretion order a quality audit of any lot of cans received and on his authority may reject lots which on audit are shown not to meet the quality criteria and limits described below. Quality complaints and decisions to reject shall be communicated by the Plant Quality-Control Manager to the Plant Purchasing Agent, who shall advise the supplier and arrange appropriate action.

V. CLASSIFICATION OF DEFECTS

    A. <u>Class A Defects</u>: faults which prevent a can from safely containing and protecting the product:

METAL CONTAINER SPECS AND QUALITY CONTROL

1. Leak anywhere on the body seam or delivered double seam
2. Seaming compound missing from any end or incomplete ring of compound
3. cracks of any size in body or ends
4. dented flange at open end, of such degree as to prevent complete double-seam formation
5. contamination of can interior with excess solder, flux, grease, rust, or any other foreign matter which is not removable in the can washer
6. missing or incomplete lining and outside enamel

B. Class B Defects: faults which make a can of borderline functionality or seriously deficient in appearance:

1. dents over 1 inch long
2 out-of-round at open end, such that closing effectiveness and efficiency may be reduced
3. seaming compound below minimum weight
4. solder bead at open end of body seam, such that cutthrough may occur in closing
5. seam stripe missing or incomplete
6. excess compound in ends, causing jams in feeding from stack

C. Class C Defects: faults which adversely affect appearance, but not function of cans

1. dents less than 1 inch long
2. scratches on ends, exterior surface

VI. ACCEPTABLE QUALITY LEVELS: For purposes of this specification, a lot shall be defined as a truckload of cans as delivered to an XYZ Plant. This will usually consist of 18 pallets holding 1680 cans each, with one pallet of loose ends, amounting to 30,240 cans and 30,240 ends. The required quality levels for each class of defect shall be as follows:

Class A	1.0% maximum defectives
Class B	2.5% maximum defectives
Class C	4.0% maximum defectives

Should any lot be audited for quality, the audit shall be made by sampling 315 specimens, 18 from each pallet including all tiers. Inspection of the sample shall be made in accordance with the schedule of defects classified in V, above. Decision to accept or reject shall be based on the number of defectives found in the sample, as follows:

Class of defectives in sample	Accept lot if	Reject lot if
A	7 or less	8 or more
B	14 or less	15 or more
C	21 or less	22 or more

It is advisable before sampling by the above plan and making formal audit to inspect for dented flanges on cans at the periphery of each pallet, since this particular Class A defect is more likely to occur on the outside of the pallet as a result of shipping conditons or loose pallet wrapping. If such a condition is found, rejection should be made on a pallet basis, rather than the entire lot.

METAL CONTAINER SPECS AND QUALITY CONTROL

PACKAGING MATERIAL SPECIFICATION
PROPERTY OF XYZ COMPANY

DATE Nov. 1, 1984   No. 520810

ITEM 20-Ounce Bean Soup Can

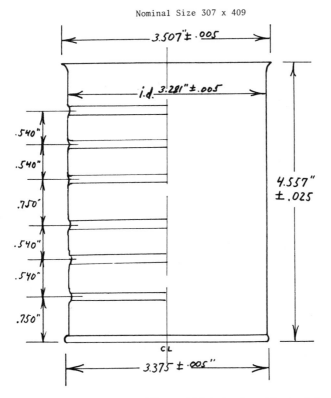

Alternative constructions: 90TU or 90T5 body plate if unbeaded, or
                            80TU or 80T5   "   "   with 6 beads

When beaded, bead depth .032 to .042". Bead width 0.250, incl. radii.

Body plate 0.25 ETP if soldered, or CCO/TFS if welded.

Epoxy enamel lining and coating, with seam stripe.

Figure 10 Sample specification: 20-ounce bean soup can.

PACKAGING MATERIAL SPECIFICATION

PROPERTY OF XYZ COMPANY

DATE Nov. 1, 1984   No. 520810

ITEM Can End, 20-Ounce Bean Soup

1. PROFILE

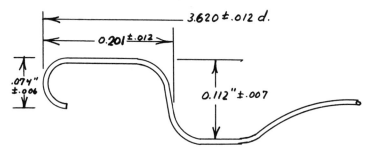

Nominal size 307

2. CONSTRUCTION

   Plate: 90T3, CCO/TFS food-grade epoxy lining & outside enamel.

   Stack height: 26 to 30 ends per inch

3. Hook detail

Figure 11 Sample specification: Can end, 20-ounce bean soup can.

METAL CONTAINER SPECS AND QUALITY CONTROL

SPECIFICATION FOR: <u>Label, 20-Ounce Soup Can</u> **NUMBER:** <u>520811</u>
PROPERTY OF THE XYZ COMPANY **EFF. DATE:** <u>November 1, 1984</u>

I. SCOPE : This specification describes the construction and performance requirements for wraparound labels to be applied on 307 X 409 cans. The corresponding can specification is No. 520810, dated November 1, 1984.

II. CONSTRUCTION: Labels shall be made of 60-pound stock, litho-coated one side.

Dimensions shall be 11 inches by 4 1/4 inches, plus or minus 1/32 inch.

At the left end of the label there shall be a vertical band unprinted (except for identification of supplier and Design No.) 7/16 inch wide, and an unvarnished band 9/16 inch wide, for lap adhesion. See Print No. 456,* which is part of this specification.

Grain direction of the paper shall be in the long dimension of the label. Printing may be done by any process which satisifes the need to match the original mechanical art. A water- and scuff-resistant lacquer shall be applied over the inks.

III. PERFORMANCE: Labels shall be suitable for application in any roll-through labeler that feeds from stacks. Stacked labels shall not stick together nor curl when stored at 40 to 65% humidity.

From this point on, the spec can be identical to No. 714753 in the previous chapter, which includes a spec for label on a 1-quart glass jar.

---

*See Figure 12.

SPECIFICATION FOR: Shipper, 12/20-Ounce Soup Cans **NUMBER:** 520812
PROPERTY OF THE XYZ Company   **EFF. DATE:** October 1, 1984

I. SCOPE: This specification covers construction and performance requirements for a corrugated box used to distribute 12 cans of 20-ounce XYZ Soups. The related can specification is No. 520810, dated November 1, 1984.

II. CONSTRUCTION: The box shall be ELC style, 4 X 3 X 1 arrangement, made of 200T board (both liners 42-pound natural kraft board), with C flutes, vertical direction. The manufacturer's joint shall be glued. Top and bottom surfaces of the box shall be treated with nonskid coating to improve pallet stability. Inside dimensions shall be 13 7/16 X 10 1/16 X 4 7/16 inches, with tolerances plus or minus 1/16 inch.

III. PERFORMANCE: Boxes shall be delivered with consistent quality for efficient erection and can loading on automatic end-loading case equipment.

IV. DELIVERY: Boxes shall be fabricated by suppliers with manufacturer's seam glued, edges scored, and KDF. The flat cases shall be stacked in bundles of 50 and tied. Stacks shall be made with the same side up on all cases, and the same edge of each at one side of the stack, for automatic erection and proper placement of the date-marking stamp.

V. INSPECTION: Suppliers shall maintain sufficient quality control to assure compliance with this specification, including responsibility for disposition of damage by his carrier in making deliveries to XYZ Plants.

Quality defects found by XYZ personnel shall be reported through the Plant Quality-Control Manager to the Plant Purchasing Agent, who will arrange appropriate action with the supplier.

If visual observation and performance of cases on a packing line indicate that a lot may be out of compliance with this specification, the Plant Quality-Control Manager may sample and inspect the lot according to the schedule in VII, below.

VI. CLASSIFICATION OF DEFECTS

   A. Class A Defects: faults which preclude shippers from protecting or identifying the product contained through the distribution to retailers and institutional buyers:

METAL CONTAINER SPECS AND QUALITY CONTROL

   1. loose manufacturer's joint
   2. dimensions outside of limits
   3. break or cut-through a liner at a score
   4. tears, punctures, holes, or ragged flaps with adhering pieces of excess board
   5. wrong copy, missing or illegible copy, or wrong color graphics
   6. contamination with any foreign matter other than fiberboard dust

B. <u>Class B Defects</u>: faults which make function borderline or questionable:
   1. incomplete or partially illegible printing
   2. off-square gluing of manufacturer's joint
   3. bends other than on score lines
   4. liner incompletely glued to corrugated medium
   5. gap greater than 1/8 inch when flaps are butted at the ends in case sealing
   6. nonskid treatment missing

C. <u>Class C Defects</u>: faults which impair appearance, but not function:
   1. light or blotchy printing
   2. stains, scratches, or scuff marks

VII. <u>ACCEPTABLE QUALITY LEVELS</u>: For purposes of inspection and acceptance, a <u>lot</u> of corrugated cases shall be a <u>truckload</u> as delivered. An average delivery will be 100 bundles of 50 cases, or 5000 cases. Inspection shall be made when necessary or advisable by sampling in accordance with MIL-STD-105D, Single Sampling, Normal Inspection. In this case, the sampling shall consist of 200 specimens, take two from each bundle, with inspection for the defects listed in VI and action taken as indicated by the following table:

Class of defects	AQL	Reject level
Class A	1.0	6 or more defectives
Class B	4.0	15 or more defectives
Class C	6.5	22 or more defectives

PACKAGING MATERIAL SPECIFICATION

PROPERTY OF XYZ COMPANY

DATE Nov. 1, 1984   No. 520811

ITEM Label, 20-Ounce Soup Can

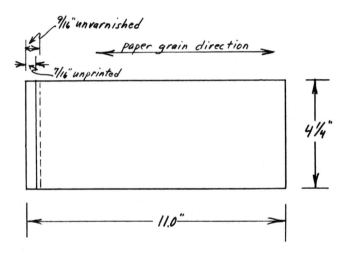

All dimensions plus or minus 1/32"

Material: 60-pound label stock, litho coated one side

Figure 12 Sample specification: Label, 20-ounce soup can.

METAL CONTAINER SPECS AND QUALITY CONTROL

I. COMMENTS ON THE SPECIFICATIONS

The reader should not accept literally all details of the can spec as being correct for any bean soup. While the product as a category is mildly corrosive, formulations can vary greatly in salt and spice concentrations and differ considerably in pH. Some can be provided with greatest shelf life by a combination of enameled ends and 1/2-pound tinplate body, not enameled, for example.

An alterative body spec is shown, with six beads and lighter steel. The inclusion of beads stiffens a can body and increase its resistance to denting. This allows plate weight reduction with dent resistance equal to that of a heavier plate without beads.

The effects on product flavor of lining enamels and seaming compounds can be determined only by storage and taste test. The spec might have equally well been written with detailed formulas for both enamel and seaming compound, although that approach leaves less flexibility for the supplier and has the risk of freezing the spec out of newer and better developments without rewriting.

Dimensional tolerances of metal containers are considerably closer than those of glass. Glass diameters at the finish can be controlled up to 0.020 to 0.030 inch and the body diameters to plus or minus 0.060 inch, while cans will vary only 0.005 inch to 0.010 inch. Jar heights will vary as much as body diameters, while can heights will vary half as much: plus or minus 0.030 inch. The reasons lie in the fact that glass containers are formed directly from a vitreous material, while cans are converted from sheet stock, and that glass containers are formed in a series of moulds (per lot), while cans are formed on a single body maker (per lot), which eliminates mould variability.

To one familiar at all with can structure, the list of defects will be self-explanatory; to others, it is suggested that one or two of the appropriate references listed in the bibliography at the end of this book will be informative.

The sampling plan specified in VI can be seen in Table 5, Chapter 4. AQL levels selected for this spec are arbitrary, and not necessarily ideal or typical, but they are used to illustrate an objective means of quality evaluation which can be adjusted to any level of tightness desired.

Of all Class A defects, dented flanges at the open end are by far the most common, which is the reason for the last paragraph of the can spec. Much of the flange damage occurs in shipment, when pallets jostle one another, and is therefore the most serious defect for which the supplier cannot inspect accurately at his plant. A superficial visual inspection for dented flanges on receipt of a lot of cans is always advisable, as a minimum effort.

The inside dimensions of can shippers are always a bit smaller than the cluster of collated cans, to prevent the cans from shifting in the case, overriding the double seams of a adjacent cans and causing dents or label tears. If specified 1/16 inch less than the cans measure, the case dimensions at their minimum would be 1/8 inch less, and at maximum equal to the can cluster. If specified any smaller, cut-throughs at the edges of the case would occur, and if larger, cans would be packed too loose.

II. QUALITY CONTROL OF CANS

Suppliers must first be concerned with the control of weight and temper in the plate they receive as a raw material for can-making. While large quantities, such as 200 or more base boxes, will vary in weight only 2 1/2% plus or minus nominal weight, individual sheets (standard size 20 $\times$ 14 inches) can vary plus or minus 10% from nominal weight. Temper or hardness variability cannot be expressed as a percentage, but Rockwell hardness ranges overlap from one temper to the next:

Temper designation	30-T Rockwell range
T-3	54 to 60
T-4	58 to 64
T-5	62 to 68

The three temper designations above are the most commonly used for can-making. Now let us see what can happen in the extreme to a can such as the No. 2 bean soup container we have just specified. Without beads, we have specified 90T-5 as one of the body-plate options; 200 base boxes of that plate will make about 100,000 bodies, and they will on the average differ from other lots of 200 base boxes only within the range of 87.5 to 92.5 pounds per base box. A single sheet will make only 4.45 can bodies, however, and groups of four will differ from other groups of four within the range of 81 to 99 pounds per base box. Thus, small groups of cans made from 90-pound plate can measure close to nominal 80-pound plate or to 100-pound plate.

It can be noted also from the table above that the low end of hardness in the T-5 range overlaps into the T-4 designation. Overall, then, the weakest can that is likely to be made from 90T-5 plate could analyze as 80T-4, and the strongest could measure up to a nominal 100T-6 (although not shown, the T-6 hardness range starts at 67).

Of course, it would be rarely that low temper coincides with low weight, but it does happen, and if a group of cans appears unusually prone to denting or collapse under vacuum, low weight and/or temper would be the factors to look for.

When the packaging specialist is designing a can spec, it is important for him to consider the performance of those cans which will appear at the low end of the plate ranges.

The occurrence of minimum and maximum weight sheets can be prevented from carrying through into cans by the latest improvements in coil-cutting lines, which can be set up to classify sheets into *low*, *nominal*, and *heavy* stacks. This eliminates the greatest extremes at can plants where such equipment is available and economical to use; that is, in large plants where many different specs are fabricated, and the sorted light and heavy sheets can readily be used without shipment elsewhere. This plate-sorting process is the first step in can quality control and is automatic, operated by very accurate mechanical or beta-ray gauges.

The next process steps are the parallel operations of body-forming and end-stamping. The body is made by cutting a "blank," which is a rectangle roughly equal in length to the circumference of the body to be made, and in width equal to the height of the body. This "blank" is wrapped around a horizontal steel tube whose diameter is the inside can diameter and is pushed along the tube of the body-maker, either across a solder bath, which solders the body seam securely if tinplate, or across a welder, if tin-free steel. The operation is done at such speed that several hundred cylinders per minute come off the end of the body former into a flanger, which rolls an outward flare onto both ends of the cylinder so that can ends may be applied.

Up to this point, the following defects can occur; usual causes are noted:

1. Crack in the body plate at the hook of the soldered-can seam—will open a leak beside the soldered seam—rare type defect—occurs only if plate is well above temper limit.
2. Leak at top or bottom end of soldered body seam—spotwelds too tight, preventing solder penetration of seam—occurs in a rash of defectives rather than a random pattern.
3. Skewed can, or one end diameter oversize—loose weld, allowing seam to spread—will occur frequently if welder needs maintenance.
4. Solder on inside of body seam, especially solder bead at one end of the body seam—maladjustment of solder roll position.
5. Skips in solder (other than Item 2, above)—solder bath too cold.
6. Cracked flange—plate temper too high or solder-horn conveyor chain too tight—Items 5 and 6 result in leakers. These defects will tend to cluster rather than appear randomly.

When this operation is running at 400 per minute, and the can buyer establishes a 1% maximum on Class A defects (the above defects lead to leakers or contamination), it requires sampling a specimen every 15 seconds to even look at 1% of the body-maker output. This is obviously impractical, so recourse is taken to automatic process control, such as temperature controllers on the solder bath, and a leak tester after application of one end, so that defectives from 1, 2, 5, and 6 above are automatically rejected at the end of the line.

To return to the manufacturing controls, let us look briefly at the end-forming process, which goes on parallel to the body-forming. The enameled plate, in sheet form or coil form, is fed into a punch press, from which stamped circles are drawn into the desired profile, rolled at the edge to form a hook that will engage the body flange, and lined with seaming compound. There are only three common defects arising from this operation: wavy seaming panel, incomplete or underweight compound application, and cracks in the plate from drawing—all lead to leakers. An uncommon defect is heavy compound application, which leads to stacked ends sticking together. If this causes two to feed at once into the double seamer, a time-consuming jam will result.

Compound is usually applied by a single nozzle into the seaming panel, distributed over two rotations of the end. Beside a 360° overlap, there will be an opening (of the nozzle) overlap, and a closing overlap, such that compound weight will not be completely uniform around the 360°. The minimum-weight spot must be heavy enough to seal the step at the crossover of the double seam—that point where the double seam crosses the notched end of the body seam. *Total* weight of compound applied is easy to audit by weighing a small group of ends before and after compounding, but local weight can vary within 45° segments by 40% plus or minus the 360° average. This variability can be tested crudely by a micrometer gauge that measures compound later thickness, or by a microtome with a blade that scrapes a section of compound out of the seaming panel for weighing.

The vacuum tester at the end of the whole can line will sort out *fast* leakers caused by any body or end or double-seam integrity defect on the combined body with one end attached, just before palletizing for shipment. Small leaks must be detected by a Maede test, which consists of pressurizing a can under water; if there is leak of any but microscopic size, it will cause visible bubbling.

The loose ends sent to the buyer must be separately inspected for defects, principally cracks and insufficient compound; 1% of these defects will produce 1% leakers, so the sampling must be adequate for control at considerably lower levels, recognizing that there will still be defectives among the cans, even after sorting with the vacuum tester.

If it were decided to hold loose-end defectives to 0.1%, then according to Table 5, the inspection of a lot of 30,000 ends calls for a sample of 315 specimens, with acceptance of the lot on finding one defective or none, and rejection on finding two or more. This is very stringent, but it must be realized that the criteria for acceptance cannot be loose if we wish to maintain an overall Class A defective level not over 1.0%. We could loosen up a bit on the end acceptance to 0.65%, in which case we would accept the lot up to seven defectives in the sample of 315, and reject on eight or more. The risk is that the vacuum tester may not eliminate all but 0.35% leakers from the assembled bodies with one end, and, of course, there are other Class A defects which must be included to make up the total of 1% (contamination and dented flanges) defectives.

MIL-STD-105D has AQLs between 0.1 and 0.65, which can be used by the quality-control specialist as experience with many lots of cans provides data on the actual performance levels of cans and loose ends with respect to leakers. For instance, 30 lots may show the following performance, or *process average*:

Contamination	0.1%
Dented flanges	0.5%
Leakers, cans, 1 end attached	0.3%
Allowable leakers, loose ends	0.1%
Total, allowable Class A defects	1.0%

In this example, the dented flanges are likely to cause leakers when the buyer fills the cans and applies the second end to close the cans, but in addition, his own process of closing will inevitably have variability which will contribute some increment of leakers. Since this would generally arise from shipping damage, improved packing of the pallets, or change of the trailer size, or the pallet arrangement in the trailer, addition of dunnage between pallets, etc., are possible steps. The size of the effort to reduce defectives should be based on two factors: the danger inherent in leakers and the cost of leaker reduction, rather than the challenge of an intellectual puzzle.

One other point of interest in the can spec should be noted: it concerns the body variation with six beads. Beading is done at the same time as flanging, that is, immediately after soldering the body seam. Leaks can be caused in beading if the solder is too hard; it will crack where the bead crosses the body seam. There are several ways of preventing or correcting leakers from solder cracks:

METAL CONTAINER SPECS AND QUALITY CONTROL

```
                      PLANT CONTROL LABORATORY

            FINISHED GOODS LOT INSPECTION - PACKAGE DEFECTS AUDIT

      PRODUCT  Bean Soup, 20-oz   PRODUCT SPEC NO. 520000    DATE STAMP 2 26 85
      PRODUCT LOT NO.  450        PACKAGE SPEC NO. 520810,-1 PACKAGE LOT NO. 713
      LOT SIZE  30,000 cans       SAMPLE SIZE  315 cans
                                    DEFECT COUNT

              Class A                   Class B              Class C
        1 defect 1 defective      1 defect 1 defective  5 defects 1 defective

      Leaks                       Dents ≥ 1"            Dents < 1"
        body seam_____          No seam stripe_____  Scratches_____
        bead x-over_____
        mfr's dbl seam_____     Excess compound_____  Label Class C_____
        XYZ    "     "
        end 1 (rec'd on)_____    Cutover_____
        end 2 (XYZ app'd)_____
                                  Label Class B_____
      Contamination
        solder, flux_____
        grease, other_____

      Incompl. enamel_____

      No date stamp_____

      Label Class A_____

      Total Class A  _____      Total Class B _____   Total Class C _____
                                                                       ──
                                                                        5
      Defectives     _____                 _____                 _____

      Limit             7                     14                     21


      Action        Accept_____     Reject_____    Re-inspect and sort_____

      Comments _____

      _____  by  _____

                                                  Date _____
```

Figure 13 Canned goods inspection form.

1. Increase the percentage of tin in the solder, which makes it more flexible, but also more expensive.
2. Be sure on the body-maker to wipe off excess solder—a thin layer is more flexible than a thick layer.
3. Place laps rather than lock construction at the body seam where the beads cross. The bead bends only two thicknesses of plate rather than four, if it crosses the seam at a lap joint.

Most of what has been said above relates to the quality of the cans as received and an understanding of defects arising from can manufacture and delivery. The packer's responsibility with respect to defectives concerns handling and closing the cans. His contribution to leakers will occur in denting flanges during the depalletizing, washing, and filling, and in maladjustments of the closing equipment. Beyond that, denting will be the most commonly occurring defect.

The control of double seaming at the closer requires the services of a specialist to keep the machinery in adjustment, to handle variability within spec limits of can lots, to replace worn seaming rolls when necessary, and to adapt the closer from the 80-pound beaded bodies to the 90-pound unbeaded bodies, if lots of these alternative specs are received from two or more suppliers. The closer specialist uses saws, files, magnifiers, and gauges to inspect seams right out of the closer. Here as much as anywhere in the packaging field, quality is controlled on the spot, not audited by inspection after the fact.

Most packers of canned goods also stamp a packing date into the loose end at its center panel as it is fed from the stack into the closer and list a missing date stamp as a Class A defect. A sample inspection form is shown in Figure 13. The quality control of labels for cans is essentially the same as described for glass container labels in the previous chapter. Shipping-container quality control will be described in Chapter 8.

7

Plastics Package Specifications and Quality Control

There are four major forms of plastics packages:

1. Film packages, such as bags, pouches, envelopes, overwraps, and shrink wraps
2. Thermoformed packages, such as trays, tubs, cups, blisters
3. Injection-moulded containers, such as boxes, tubs, and cups, and components such as closures and lids, sometimes combined with extruded tubing
4. Blow-moulded bottles

In addition, and in lesser volume than any of the above, compression-moulded packages and components comprise a fifth category of plastics. Phenolic and urea-formaldehyde closures for cosmetic and chemical packages and expanded polystyrene beads moulded into cushioning trays are the greatest part of the packaging products made by the compression-moulding process.

Plastic films and packages made from them are covered in Chapter 8. Compression-moulded products will not be covered separately in this book, in our discussions on specs and q.c., since the process technology is quite similar to that of injection moulding, and the product quality criteria and classification of defects also resemble those of injection-moulded parts.

This chapter, then, will present sample specs for injection-moulded, blow-moulded, and thermoformed packaging elements; common defects will be classified, and comments on possible quality control actions will be made.

I. BLOW-MOULDING AND INJECTION-MOULDING SPECIFICATIONS

A good example that will cover both blow-moulded and injection-moulded part specs is a plastic shampoo or bath oil bottle, with a thermoplastic snap cap. This example is not a spec that describes exactly any package now on the market, nor one which ever was, but like the specs described in the two previous chapters, it is illustrative of the elements which should and can be included in any good spec. In this instance, we shall postulate a 1-pint squeeze bottle for bath oil, made of pigmented low-density polyethylene, and closed with a beaded polypropylene cap. The decoration is silk-screened directly onto the bottle.

PLASTICS PACKAGE SPECS AND QUALITY CONTROL

SPECIFICATION FOR: 1-Pint Bath Oil Bottle   NUMBER: 945600
PROPERTY OF THE XYZ COMPANY   EFF. DATE: January 15, 1982

I. SCOPE: This specification describes construction and performance requirements for a plastic blow-moulded squeeze bottle to contain and dispense XYZ Bath Oil. It applies to Plants 11, 17, and 19 and is not a replacement for any prior spec.

II. CONSTRUCTION

   A. Bottles shall be made of low-density polyethylene in the range of 0.92 to 0.925, pigmented in opaque yellow-orange color in accordance with a standard identified as PRX-4337. Resin shall include a slip agent.

   B. Weight of bottles as delivered shall be in the range of 32.0 to 33.0 grams.

   C. Dimensions and capacity shall comply with Print No. 116,* which is part of this specification.

   D. Wall thickness distribution shall be sufficiently uniform to withstand normal stresses in filling, distribution, and use. It is desired that no point on a bottle have a wall less than 0.010 inch, to minimize the hazard of failure and leakage.

   E. Decoration shall be silk-screened directly onto the label area, using graphics which are separately specified. The decoration shall be fast to the oil contained and to soaps, detergents, hot water, and abrasion.

   F. Bottles shall stand without rocking, and the bottom pinchoff shall be recessed above the plane of the standing surface.

   G. Bottles shall be free as commercially possible of visual defects, such as streaky color, embedded foreign particles, rough spots from non-uniform mould surfaces.

   H. Leaners shall not exceed 1/8 inch on a side.

III. PERFORMANCE

   A. Shelf life of XYZ Bath Oil in the bottles described herein shall be not less than 1 year, when evaluated in accordance with Research

---

*See Figure 14.

Department methods against their standards, which shall be made available to prospective suppliers on the authority of the General Purchasing Agent.

B. Distortion resistance of filled bottles under vertical load shall be a maximum of 1/8 inch sag with 6-pound dead load for 6 weeks at 80°F. Bottles in cases (Spec No. 945602) shall be capable of warehouse stacking up to a height of 21 feet for 12 weeks at 80°F and 50% r.h. without more than 1/8 inch height distortion of bottles in the bottom layer.

IV. <u>MANNER OF DELIVERY</u>: Bottles shall be bulk packed for delivery in corrugated boxes with polyethylene bag liners, 250 bottles per box. Boxes shall be closed with paper tape only. If convenient, local arrangements may be made between XYZ plants and their suppliers for return and reuse of the boxes and bag liners.

All trim waste shall be separated from bottles before filling them into the lined boxes, to assure that plastic trim will not get into XYZ packing-line equipment. Boxes may be grouped on pallets for delivery to XYZ plants, but such pallets may only be 48 X 40 inches, to avoid problems with another size being accidentally mixed with XYZ standard 48 X 40s.

V. <u>INSPECTION</u>: A lot shall be defined as a truckload, usually consisting of 176 boxes or 44,000 bottles. The supplier shall certify that each lot has been inspected before shipment, and that it meets the following quality criteria:

Class of defect	AQL
A	0.1
B	1.0
C	2.5

Should it become necessary as a result of observed defects in XYZ production to make an audit of lot quality, a sample of 500 bottles shall be taken randomly, but with all cavities represented. Inspection of the sample shall follow the classification of defects listed below, with acceptance/rejection on the following criteria:

| | Number of defectives ||
Class of defect	Accept	Reject
A	1	2
B	10	11
C	21	22

Unsatisfactory packing-line performance shall be reported to the Plant Purchasing Agent, if found related to identifiable defects. He will arrange appropriate action with the supplier to sort or replace the lot in question if defectives exceed the AQL.

VI. CLASSIFICATION OF DEFECTS

A. <u>Class A</u>: faults related to possible breakage and failure to safely contain the product during storage, distribution, and use:
   1. cracks and breaks anywhere on the bottle
   2. any spot at the heel or shoulder less than 0.005 inch in thickness
   3. rough cutoff at finish, leading to leaky closure

B. <u>Class B</u>: faults related to poor packing-line performance or poor resistance to warehousing stresses:
   1. leaners in excess of 1/8 inch
   2. below minimum weight
   3. bottom rockers from insufficient recess of pinchoff, or adhering trim waste
   4. any dimension outside of limits shown on Print No. 116

C. <u>Class C</u>: defects in appearance or decoration:
   1. blotchy or incomplete printing
   2. wall thickness less than 0.010 inch, such as to lose opacity
   3. streaky pigmentation from inadequate pigment blending

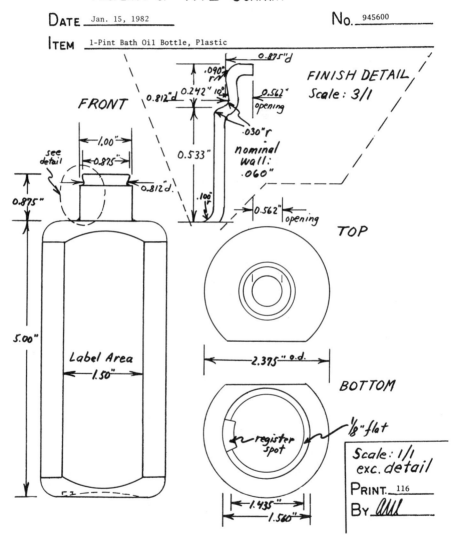

Figure 14 Sample specifications: 1-pint bath oil bottle, plastic.

PLASTICS PACKAGE SPECS AND QUALITY CONTROL

SPECIFICATION FOR: <u>Plastic Closure, 1-Pint Bath Oil</u> NUMBER: <u>945601</u>
PROPERTY OF THE XYZ COMPANY   EFF. DATE: <u>January 15, 1982</u>

I. <u>SCOPE</u>: This specification describes requirements for an injection-moulded polypropylene closure to fit Spec No. 945600, blow-moulded polyethylene bottle. It is for use in Plants 11, 17, and 19 and is not a replacement for any prior spec.

II. <u>CONSTRUCTION</u>: The closure shall be a snap-fit ring with a plug-fit aperture and captive plug, injection-moulded in one piece of polypropylene. The closure shall make a liquid-tight fit to the trimmed bottle finish, and the plug shall also make a liquid-tight fit with the central aperture of the closure.

  A. Polypropylene resin shall be resistant to stress-cracking from contact with XYZ Bath Oil for not less than 18 months at $70°F$. Grange Petrochem. Co. No. 5545 Resin or its equivalent are acceptable.

  B. Dimensions of the closure shall comply with Print No. 117,* which is part of the specification.

  C. Closures shall be made from resin compounded to Color No. PRX-4311, a standard chip of which will be supplied by the General Purchasing Agent to prospective suppliers. This color is a midnight blue, intended to match the ink used in silk-screen decoration on the bottle.

III. <u>DELIVERY</u>: Closures shall be packed in bulk in polyethylene-bag lined corrugated boxes, 1000 closures per box. All moulding flash and mould runners shall be completely separated from the closures before packing for delivery, to preclude the interference of foreign particles with capping operations and effective closure. If closure deliveries are palletized by agreement with the purchasing XYZ Plant, only 48 X 40 pallets may be used, to avoid accidental mixing of another size with this XYZ standard. Each case shall be stenciled with the supplier's lot number of numbers and the identification "PLASTIC CLOSURE—SPEC NO. 945601."

IV. <u>INSPECTION</u>: Suppliers are expected to control the quality of their products in accordance with the criteria listed below in V and the AQLs

---

*See Figure 15.

in VI. If it should become apparent to XYZ personnel in using a lot of these closures that the defective level may exceed the AQL, the Plant Quality-Control Manager may conduct an audit on that lot. If the inspection shows that such is the case, he shall advise the Plant Purchasing agent, who shall ask the supplier to sort the defectives out of the lot or replace it.

- A. For purposes of this inspection, a lot shall be defined as a shipment received, usually a truckload. If more than one manufacturer's lot is indicated on the bill of lading and the stenciling on the cases, and it becomes necessary to inspect, the supplier will be asked to bear the excess cost over inspection of the shipment as one lot.
- B. Each cavity of the injection mould shall have engraved in it an unique cavity number, and if the supplier has more than one mould, the mould number shall appear on each cap in addition, or there shall be no duplication in cavity numbers, regardless of the number of moulds.
- C. Depending on size of truck and plant release, a lot may vary between 200,000 and 400,000 closures. When sampled as one lot, in accordance with MIL-STD-105D, the sample size is 800 specimens. These shall be taken equally from each of the 200 to 400 boxes in which delivered, and all cavity numbers shall be represented.

V. CLASSIFICATION OF DEFECTS

- A. Class A: faults related to leakage or failure to safely contain the product for 1 year:
    1. partially-filled cavity
    2. flash or rough parting line that prevents liquid-tight closure
    3. outside of dimensional limits shown in Print No. 117
    4. weld lines that permit stress cracking in contact with product
- B. Class B: faults related to consumer acceptance:
    1. flash or rough parting line at points other than closure surfaces
    2. uneven color or off color or light spots caused by foreign matter embedded in plastic
    3. rough or orange-peel surface
- C. Class C: faults in nonfunctional properties or manner of delivery:
    1. plastic trim in boxes delivered with closures
    2. improper identification of cases in which closures are delivered
    3. boxes off count by more than 1%, or average of a shipment off by more than 0.1%

VI. <u>ACCEPTABLE QUALITY LEVELS</u>: Lots of delivered closures are expected to meet the following requirements:

Class of defect	AQL	Defect count Accept	Reject	Sample size	Lot size
A	0.1	2	3	800 caps	200M to 400M
B	1.0	14	15	800 caps	200M to 400M
C	2.5	2	3	32 boxes	200 to 280
C	2.5	3	4	50 boxes	281 to 400

Defects in Classes A and B apply to individual closures, and the inspection is therefore based on the lot as a single universe of closures. Class C defects, on the other hand, apply to boxes of closures, and the inspection is based on the lot received as a group of boxes.

During the inspection of caps for Class A and Class B defects, note shall be made of Cavity Numbers of defectives. Should it occur that a majority of defectives in a given lot come from one cavity, the sublot of all caps from that cavity shall be sorted out of the total lot at the supplier's expense.

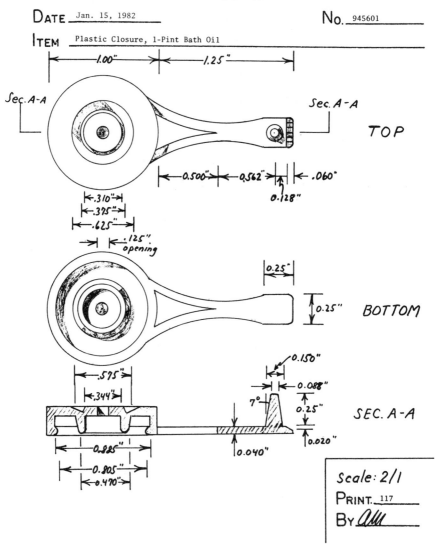

Figure 15 Sample specification: Plastic closure, 1-pint bath oil.

PLASTICS PACKAGE SPECS AND QUALITY CONTROL 141

Comments on Specifications for Blow-Moulded and Injection-Moulded Components

Since the blow-moulded bottle and the injection-moulded closure are part of the same package, it would be well to insert the comments on these two specs here, before moving on to an example of a spec for a thermoformed component.

It will be noted that the spec for the blow-moulded plastic bottle resembles that written for a glass container in an earlier chapter. There are, in fact, several parallel processes in making the two types of containers: both are blown into a set of moulds, and the variables which cause several kinds of defects are similar, such as thin spots, leaners, "rocker" bottoms, unfilled moulds from blocked vents. Plastic bottles have much thinner walls, on average, than glass, running 0.020 to 0.030 inch, while glass runs 0.050 to 0.100 inch.

The pinchoff at the bottom of a plastic bottle has no counterpart in glass, however, since the glass bottom is pressed, while the plastic bottom is cut off from an extruded tube. Control of the pinchoff in the moulding operation is very important to the determination that the bottle will stand firmly on its bottom.

In the chapter on glass specs, it was indicated that cullet is returned to the furnace and remelted. Similarly in plastic, the finish trim and the pinchoff trim are reground and reextruded, which greatly reduces the net waste in the manufacture of a lot. It is vital to good bottle quality, however, that the trim be kept free from contamination during collection, storage, and regrinding, since dust and dirt picked up from the floor, storage barrels, or grinder are the usual cause of black spots and streaks or other embedded particles in finished pieces. In the remelting of glass cullet, furnace temperatures are so high as to volatilize all organic foreign matter which may have been picked up, but such is not the case in a plastics extruder, so additional care must be taken to keep plastic regrind clean.

Dimensional control of plastics is considerably closer than that of glass, because moulding temperatures are much lower, and the moulds are consequently subjected to much less attack and wear. It will be noted in Figure 4 that major dimensions of the mayonnaise jar show tolerances of plus or minus 3/64 and 1/16 inches (0.045 to 0.062 inch) while Figure 14 shows tolerances of plus or minus 0.025 inch in height, and 0.012 inch in diameter. By and large, the dimensional variability of polyolefin plastics is 0.005 inch (plus or minus) per inch of any dimension and arises almost entirely from cooling shrinkage in the mould.

On the other hand, the tolerance shown in Figure 4 for glass is *not* of that nature, but rather the result of mould wear from the beginning to the end of its life. The high temperature of molten glass coming into contact with the inside of the mould causes pitting of the cast-iron cavity and carbonization of the lubricant with which the cavity is periodically swabbed. Each time the moulds are removed from a glass-blowing machine after completion of a lot, they must be sand-blasted to clean off the carbon deposits. This causes a slight enlargement of the cavity, so the next lot of bottles blown will be a bit larger on average than the prior lot. Thus, a set of glass moulds is usually made to the minimum of the dimensions shown on the print, and the first lot produced has dimensions near the minimum. The next lot will be a bit larger, on the average, because the cavities have been sand-blasted between making the two lots, and so on, until the moulds have been worn to a size which makes jars just at the maximum of the tolerances, after which they must be replaced.

A further complication in the life of moulds for glass is that the cavities tend to wear more and chip slightly at the parting lines. This requires periodically shaving the cavities at the parting lines, which reduces the cavity diameter at right angles of the plane of the parting line and leads to a slight out-of-round condition in jars subsequently blown. The tolerance thus shown on the glass print becomes an out-of-round tolerance as well as an average dimensional tolerance.

Figure 16 illustrates with two control charts the manner in which blown glass and plastic bottles will differ long-range in dimensional variability. The successive lots of glass blown on a given set of moulds will show:

1. a steady upward drift in the average dimension
2. an increasing lot variability as individual cavities wear more or less than others, or go more out of round

The plastic bottles show a more random pattern of variability, with some differences from lot to lot arising from slight differ-

PLASTICS PACKAGE SPECS AND QUALITY CONTROL

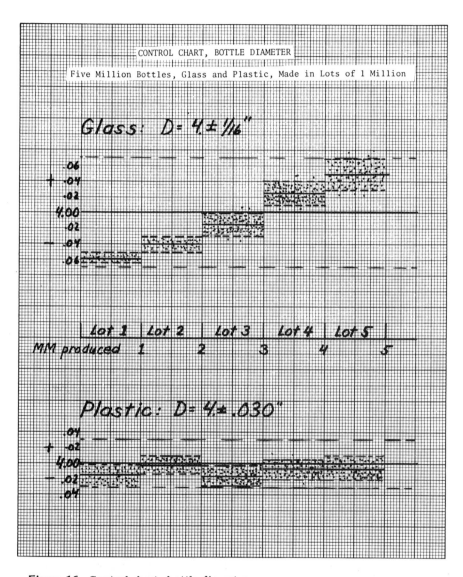

Figure 16 Control chart—bottle diameter.

ences in the raw material as to density and thermal coefficient of expansion, or from extrusion conditions, such as temperature and pressure and mould cooling rate.

Another distinct difference between glass and plastic bottles is in load-bearing ability. Well-designed and well-made glass containers need their shipping container for impact resistance, while in warehousing the glass is strong enough to carry all of the vertical load of a 20-foot stack. Just about the opposite is true of the plastic container; it has great impact resistance, but limited load-bearing strength. The shipper has to carry most of the weight in a warehousing situation. It is therefore not uncommon to see a shipper for glass with light single-wall partitions, while the equivalent for plastic bottles will have a set of double-wall partitions or a case liner.

In Spec No. 945600 for the 1-pint bath oil bottle, the weight is specified in II. CONSTRUCTION and the load-bearing strength in III. PERFORMANCE. The weight alone is not an adequate spec, since if the distribution of plastic in the bottle is poor, it could have a heavy bottom and thin side wall. We can calculate that, since the bottle is about 6 inches high, a 21-foot stack would be about 42 bottles high. A 1-pint bottle, filled, will weigh about 1 pound, so the load-bearing spec says that the bottle should be able to hold six more on top of itself. In a stack of 42, then, the shipper must have strength to hold 6/7 of the total load. Why require the bottle to carry a load of 6 pounds by itself? This permits the building of a store display, in which the shippers are cut into trays, partitions removed, and stacked on one another. A display 3 1/2 feet high will consist of seven trays in which each bottle of the bottom tray bears the weight of six bottles above it. Such retail displays are quite common, and the spec recognizes this potential activity.

For both the bottle and the closure the AQL on Class A, leakage-associated defects is very low. The reason is that leakage from any one bottle could damage more than a case of finished goods during storage and distribution. The contents of a cracked bottle could trickle down through a whole pallet of cases and render many bottles unsalable.

References in the bottle and closure specs to blotchy or uneven color relate to the process by which color is added to a batch of

PLASTICS PACKAGE SPECS AND QUALITY CONTROL 145

"natural" resin. The color starts out as a dry powdered pigment which must be uniformly dispersed throughout the batch of resin. Poor dispersion results in a low tinctorial strength (weak color) and only partial opacity. The moulder of the parts may buy the color as a powder or as a color concentrate. In the case of the former, he must feed resin and pigment into an extruder and chop the extruded plastic before feeding it into the blow-moulding or injection-moulding machines. If the latter, he can buy the color concentrate as a resin into which the color has already been added by the color supplier or an intermediate processor who specializes in color blending. This color concentrate is then fed into the moulding machine as a small percentage along with natural resin. The moulder pays more for the color concentrate than he would pay for dry color in powder form, but he saves himself the cost of blending in the dry color, plus the hazard of airborne pigment dust getting into batches of plastic where it should not be, plus the cost and problems of cleaning out the extruder and chopper after running the color batch. For reason of these problems, few moulders do their own color blending and purchase color concentrates almost exclusively.

The details of shipping case for the packed bottles of bath oil are not presented (Spec No. 945602), because most of the material would be repetitious of case specs in earlier chapters. A few comments for this particular situation are in order, however:

1. As noted above, the shipper bears most of the vertical load stresses in warehousing and distribution. The case must be designed with adequate column crush strength, which can be accomplished with test partitions or double-wall partitions or case liners, or some combination therof.
2. To assure the vertical loading is evenly distributed over a shipper, it would be best to make it a special-slotted case (full inside flaps) or to insert plug pads top and bottom between the short flaps if a regular slotted case is used. If the case contains only 12 bottles, the plug pads would be too small for easy placement and control; it would probably be less expensive to use a special slotted case and avoid the labor cost of plug pad insertion.
3. Flutes of all components—case, partitons, and liner, if any—should run in the vertical direction, of course, to develop maximum stacking strength.

II. THERMOFORMING SPECIFICATIONS

Small articles of hardware, such as tacks, screws, magnetic catches, and drawer pulls, are commonly packed in clear plastic blisters which are heat-sealed to coated paperboard rectangles for rack and other self-service merchandising. The package structure starts with two sheet materials: printed paperboard and extruded or calendered plastic sheet. The plastic sheet material is heated to softness and vacuum- or pressure-formed into a cavity or over a male mould, leaving a trimmed flange. The printed board is coated with a thin layer of the same plastic material. Product is filled into the plastic bubble or blister, the board is placed over the product, coated side against the flange of the blister, and the two are heat-sealed together. To make an effective package, the plastic blister and board coating must be transparent, the finished package should have heat seals strong enough to require tearing or cutting the paperboard to retrieve the product, and the plastic blister must be rigid enough to hold it shape but flexible and tough enough to show no brittleness in distribution.

The plastics that are suitable for the requirements slated above include:

Oriented polystyrene
Cellulose esters—acetate, propionate, and butyrate
Polyolefins—polyethylene and ionomers
Polyvinyl chloride

The brief description above does not by any means cover the whole of thermoforming, as to either processes, materials, or end uses; it simply exemplifies one kind of application. Drugs, cosmetics, toys, personal articles, tools, hobby materials, and hundreds of other categories are distributed in skin packages, curtain-coated and drape-formed packs, stretch-packs, slide packs—all being different expressions of the same thermoforming principle.

The following specification covers in detail a typical case, which is that of the preformed blister-and-card combination outlined above.

PLASTICS PACKAGE SPECS AND QUALITY CONTROL

SPECIFICATION FOR: <u>Blister Package, 2-Inch Hinges</u> **NUMBER:** <u>8223</u>
PROPERTY OF THE XYZ COMPANY **EFF. DATE:** <u>February 15, 1981</u>

I. <u>SCOPE</u>: This specification states construction and performance requirements for a thermoformed transparent blister-on-card package suitable for the distribution and self-service rack merchandising of 1 pair, XYZ 2-inch brass hinges.

II. <u>CONSTRUCTION</u>: The unit package shall be made of two components, a rectangular printed card and a clear colorless rectangular plastic blister.

    A. The two components shall be made to dimensions shown in Print No. 202, which is part of this specification.

    B. Card stock shall be 28-point (0.028 caliper), clay-coated natural kraft board. After printing, a vinyl lacquer coating shall be applied on the front or top surface of the board, as a weight of 1.5 plus or minus 0.5 pounds per 1000 square feet, for heat-seal application of the plastic blister. The vinyl coat may be applied by knife-coating or printing screen.

    C. Plastic blister shall be formed from 0.010-inch rigid vinyl sheet, by any process which maintains thickness equal to or over 0.005 inch at all parts of the formed blister. Drape or vacuum forming over a male mould is preferred, since the risk of thin spots at the exposed corners is minimized.

    D. A hole shall be die-cut at the top center of the card, 1/4 inch in diameter as shown in Print No. 202,* for rack display.

III. <u>PERFORMANCE</u>

    A. When the flange of the blister is heat-sealed to the card, the coating adhesion to the card shall be such that fiber tear is required to open the sealed package.

    B. Package integrity shall be sufficiently strong that there shall be no separation of plastic from board if filled packages are dropped from a height of 6 feet onto a smooth hard floor at any angle, side, corner, or surface.

    C. The inner and outer surfaces of the plastic blister shall be smooth and scratch-resistant, such that there shall be no visible scratching.

---

*See Figure 17.

IV. DELIVERY OF PACKAGING MATERIALS

A. Cards shall be delivered in stacks of 500, banded and boxed, 20 stacks to a box.

B. Blisters shall be delivered nested in stacks of 100, 20 stacks to a box.

C. Boxes shall be labelled with this specification number, manufacturer's name and address, lot number and date of manufacture.

V. INSPECTION:
For purposes of this specification, a lot is commonly the quantity delivered at one time, or a single date of manufacture, usually 10 boxes or 100,000 pieces. The supplier is expected to have controlled quality to the following AQLs and inspected before shipment to the XYZ Co.:

Class of defect	AQL
A	1.0
B	2.5
C	4.0

Should it appear to XYZ personnel in using any lot of cards and blisters that defectives exceed the AQLs, the Plant Quality-Control Manager may audit for quality accoridng to the following plan:

Sample Size . . . . . .   500 (Classes A,B); 315 (Class C)

Class of defect	Number of defectives found	
	Accept	Reject
A	10	11
B	21	22
C	21	22

Lots found not to meet AQLs will be rejected to the supplier(s) for sorting out defectives.

VI. CLASSIFICATION OF DEFECTS

A. Plastic blisters

1. <u>Class A</u>: faults which compromise package integrity:
    a. cracks or holes anywhere on a blister
    b. wall thickness less than 0.005 inch anywhere on a blister.
    c. draw too shallow for containment of product

PLASTICS PACKAGE SPECS AND QUALITY CONTROL 149

2. Class B: faults which reduce salability:
   a. scratches and gels, waves, fisheyes, and other sheet defects which impair transparency
   b. embedded foreign matter, discoloration, and other visual defects
   c. failure to meet scratch-resistance test, Section III.C
3. Class C: minor defects:
   a. rough-cut flanges
   b. mould marks or "drawing lines" on sides of blister

B. Cards

1. Class A: faults which limit package integrity:
   a. insufficient vinyl coating to achieve fiber-tear heat seal
   b. tears or punctures anywhere on the card
2. Class B: faults which reduce salability:
   a. illegible or smudged printing on any part of a card
   b. color off standard
   c. rough-cut edges
   d. warpage in excess of 1/16 inch from flat
   e. absence of hole for rack mounting
3. Class C: minor defects:
   a. fiber dust adhered to top surface of card
   b. spots or stains on top surface

PACKAGING MATERIAL SPECIFICATION

PROPERTY OF XYZ COMPANY

DATE 15 Feb. 1981 No. 8223

ITEM Blister Package, 2-inch Hinges

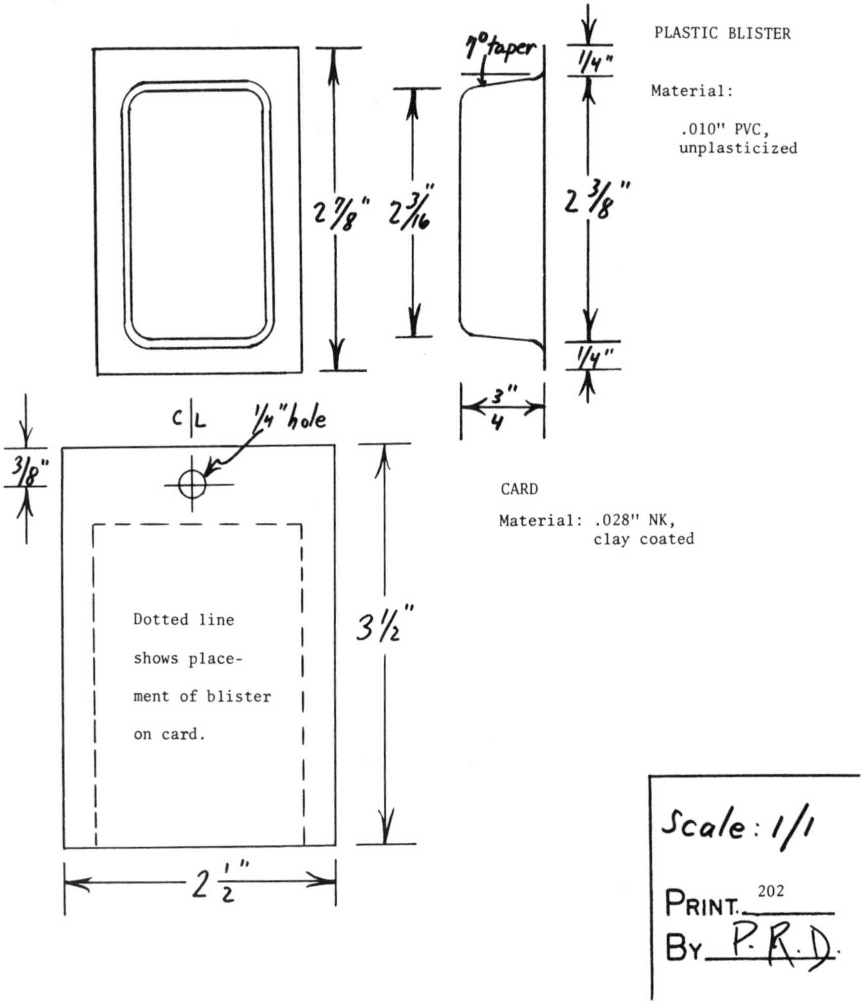

Figure 17 Sample specification: Blister package, 2-inch hinges.

PLASTICS PACKAGE SPECS AND QUALITY CONTROL

Comments on the Specifications for Thermoformed Package

The most deserving item for further explanation is the sampling plan in V. INSPECTION, where the second table calls for a sample of 500 to be examined for Class A and B defects, and only 315 for Class C. In the different samples the same number of defectives are the basis for acceptance and rejection of both B and C defects.

The reasons lie in the foundation mathematics for Table 5, Chapter 4, an excerpt from MIL-STD-105D. At AQL = 2.5, it is necessary to inspect 500 specimens to determine with reasonable accuracy whether there are 2.5% defectives in the lot of 100,000 pieces. But at AQL = 4.0, where defectives may be more numerous, it is not necessary to examine so many pieces to estimate whether the lot contains 4% defectives. Thus, the sample size is reduced to 315, that of the next smaller lot size. The same result could indeed be obtained by examining 500 specimens for Class C defects, but it would take more inspection time, and the result would be more precise than necessary. The sampling and inspection procedures in MIL-STD-105D are, in fact, a blend of theoretical and practical considerations in the measurement of quality, which, after years of work by hundreds of experts in the development and refinement of that standard, have achieved an excellent balance of great value to the packaging field as well as any other field where multiple units are made to a standard or specification.

Implicit, of course, to this specification for blister-pack materials, like all others, is another in-plant Process Specification for filling and sealing the packages. This spec will set a procedure for filling, such as: two hinges first into the blister, followed by 12 screws; then transfer the filled blister to the pocket of a heat sealer, place the card on the flanges of the blister, and seal with a given pressure, temperature, and dwell-time.

This process will have its own sampling and inspection requirements for examination of the heat seal, "squareness" of card placement on the blister, and accuracy of screw count, for instance.

8

Paper, Board, and Flexible Package Specifications and Quality Control

In this chapter will be considered a few topics that will round out the survey of specifications for representative packages made from the most commonly used materials. The use of paper and flexible material will be exemplified by the packaging of bread in a waxed-paper wrap and an outer polyethylene bag. A spec for detergent carton will cover paperboard packaging, and a final section will be devoted to the quality-control aspects of corrugated shippers.

I. BREAD PACKAGE—PAPER AND POLYETHYLENE

A waxed-paper bread wrap and an outer bag of polyethylene film seem so simple as to be trivial, but as we shall see, there are many aspects of each which are entirely essential to satisfactory performance. In comparison to the glass, metal, and plastic packages described in earlier chapters, the components of the bread package have a somewhat different relation to one another. During distribution, both the waxed paper and the plastic bag protect the bread against moisture loss. The paper wrap holds the slices in place, while the bag contributes little mechanical protection. The paper wrap carries the whole burden of communication; the bag is unprinted but must be transparent to make the wrapper visible.

When the consumer starts to use the loaf, the waxed wrapper, being opened, has lost most of both its technical functions—holding the slices and retarding moisture loss—and the plastic bag takes over both of them. The bag continues to be the reclosable protective component as long as any bread is left, while the waxed wrapper becomes less functional and more of an impediment as the number of remaining slices decreases. Thus, during the life of the loaf, the waxed wrapper starts as the most important package component and ends as the least useful, while the plastic bag is technically unnecessary at the start, but becomes more functional as the cycle proceeds.

PAPER, BOARD, AND FLEXIBLE PACKAGE SPECS   155

SPECIFICATION FOR: 2-Pound Bread Wrapper   NUMBER: 72140
PROPERTY OF THE XYZ COMPANY   EFF. DATE: March 1, 1982

I. SCOPE: This specification states the construction and performance requirements for a waxed-paper bread wrap to hold a 2-pound loaf of sliced XYZ White Bread, Spec No. 72100. It is for use in combination with Spec No. 72160, Plastic Bag for 2-Pound White Bread.

II. CONSTRUCTION

  A. Paper stock shall be 30-pound MF (machine finish) bleached sulfite, with the following properties:

		Test method
Tensile strength		
Machine direction	15 lb/inch	ASTM D828
Cross direction	7.5 lb/inch	
Tear resistance		
Machine direction	24 grams	ASTM D689
Cross direction	30 grams	

  To increase opacity when waxed, the paper shall contain the equivalent of 2% TiO_2 (0.6 pound/ream).

  B. After printing, paper shall be wet waxed with an addition of 15 plus or minus 2 pounds of food-grade paraffin wax containing 10% ethylene vinyl acetate (total waxed weight 45 plus or minus 2 pounds). The finished sheet shall be bright, glossy, and opaque defined as follows:

		Test method
Brightness	75 plus or minus 5% (GE, 460 m)	ASTM D985
Gloss	63/100	B&L Glossmeter
Opacity	50%	TAPPI T425

III. PERFORMANCE

  A. Wrapper shall be heat sealable in the range 150° to 180°F.

- B. WVT rate shall be not over 1 gram/24 hour/square meter flat (ASTM E96) nor over 2 grams creased.
- C. Roll stock shall not block at temperatures below 115°F nor pick, below 110°F.
- D. It is intended that the applied wrapper with end labels, and without polyethylene bag, shall protect the loaf contained from loss of moisture, such that when stored at 70°F and 30% r.h., weight loss shall not exceed 10 grams per day over a 3-day period.
- E. Mechanical strength of a properly applied wrap shall be resistant to rupture, such that a loaf may be dropped onto each of its six faces from a height of 30 inches to a hard smooth surface, without tearing or bursting any part of the wrap, and without failure of any seals.

IV. <u>MANNER OF DELIVERY</u>: Bread wrapper shall be delivered in roll form on 3-inch cores, 1000 print repeats per roll. Roll width shall be 18 inches, plus or minus 1/16 inch, and repeat length shall be 18 inches plus or minus 3/16 inch, as measured by distance between printed electric-eye spots.

Each roll shall be free from foreign matter and protected in shipment by a polyethylene bag or film wrap which must cover both ends as well as the outside of the roll. Outside the polyethylene shall be an overwrap of single-face corrugated or solid board, fastened with two plastic straps.

The following information shall be stenciled or clearly labeled on each roll: this specification number, lot number or date of manufacture, supplier. Also, any roll that consists of two or more spliced sections shall be clearly stenciled "spliced." No short rolls or unspliced sectional rolls will be accepted. Splices shall be heat-sealed in register.

V. <u>CLASSIFICATION OF DEFECTS</u>: Following is a list of critical, major, and minor defects. Any roll or wrap found with critical defects will be rejected. Rolls with major defects will be set aside for inspection and action as noted in VI, below. Rolls with minor defects will be used, with a record kept, and notification to the supplier.

- A. Critical defects: faults which prevent use of a roll, or which result in failure to protect the bread with an integral sanitary wrapper:
  1. crushed roll ends or core, such that a roll cannot be mounted on or fed to an overwrapping machine
  2. dirt contamination
  3. tears or holes

PAPER, BOARD, AND FLEXIBLE PACKAGE SPECS

       4. skips in wax coating or blocking
       5. no splice between pieces in a roll

  B. Major defects: faults which impair product protection, communication to the consumer, or machinability:
       1. dimensions outside of specified tolerances
       2. skips in wax coating or coating weight below minimum of tolerance
       3. missing color or illegible printing
       4. wax picking
       5. damaged sides of rolls, such that wrinkling impedes feeding off the roll, or tracking of electric-eye spots

  C. Minor defects: faults which impair appearance, but not function:
       1. gloss and/or opacity below minimum
       2. blotchy or rough printing
       3. misregister more than 0.015 inch, when printed more than one color
       4. wrinkles or folds found in roll stock as delivered
       5. inks off color, or outside of light-and-dark tolerances

VI. <u>INSPECTION</u>: Since it is impossible to execute nondestructive representative sampling on roll stock, it is more than usually important that the supplier maintain adequate quality control to assure compliance with this specification. Acceptance/rejection shall be made on an individual roll basis at XYZ plants. Supplier(s) shall identify each roll by stenciling on its kraft outer wrapper:

    Supplier's name and plant, if more than one supplies
    Date produced, and supplier's lot number
    Serial number of the roll, and number of print repeats
    Number of splices, if any

XYZ wrapping-machine operators shall be trained to identify the defects defined in V, above. When starting a fresh roll, the operator shall remove the stenciled area from the wrapper and hang it beside the stand until the roll is used. He shall remove the first two whole repeats from the starting end of the roll and staple it to the stenciled area of the wrapper.

Should the operator find a Class A defect as he runs the roll, he shall set it aside, with its identification, and start another roll. Defects of Classes B and C shall be noted on the identification sheet. If Class B defects seriously impair production or, in the opinion of the operator, may be unsatisfactory for product protection or communication, he may set the roll aside with its identification for decision by a Quality Inspector.

At the end of a shift, each operator shall turn over to a Quality Inspector the samples and identification sheets of all rolls which he has run out, and list those which he has set aside for Class A or B defects.

VII. <u>ALLOWABLE DEFECT LEVELS</u>

    A. Class A (critical) defects: Any roll with one or more defects shall be rejected to the supplier.

    B. Class B (major) defects: Out-of-tolerance dimensions, damaged rolls, and wax picking will cause enough production problems to require rejection of a roll. If the operator, in sampling the starting end of a fresh roll, considers the web to be light in weight, he may hold the roll aside until tested for wax coating weight. If found to be under the minimum, the roll shall be rejected.

    C. Class C (minor) defects: When a roll is run, it shall be rejected for minor defects if more than three repeats per hundred are blotchy, wrinkled, or off-color, or obviously out of register when printed more than one color.

PAPER, BOARD, AND FLEXIBLE PACKAGE SPECS

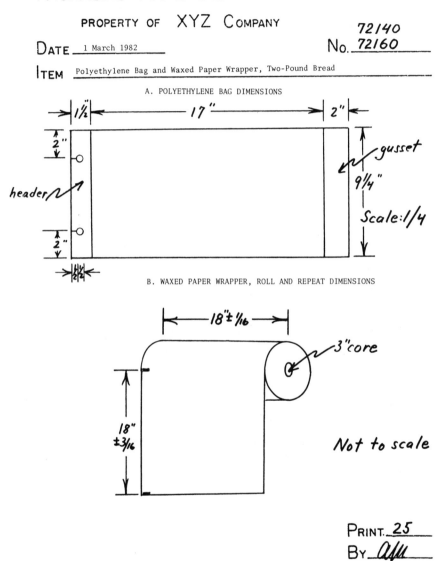

Figure 18 Sample specification: Polyethylene bag and waxed paper wrapper, 2-pound bread.

Comment on the Specification

This spec differs greatly from any of those described in earlier chapters, for reasons noted in Chapter 4 on the example concerned with roll stock. By their very nature, materials in roll form cannot be systematically sampled without coming into direct conflict with common-sense economics and production efficiencies. A few specifics will illustrate:

1. Sampling any spot other than the outer end of a roll would require rerolling it and causing splices—creating defects.
2. The operator of the wrapping machine must also be a trained inspector—it would be unjustifiable to station a Quality Inspector beside each machine full time.
3. Other than those defects which can be observed on the outside of the roll, such as a crushed core or damaged sides, defects of all classes can occur randomly anywhere inside and cannot be anticipated at any particular point. Only statistical sampling or 100% inspection could reveal these defects reliably, and 100% inspection occurs only at the time of use in production.

These conditions therefore require that the acceptance/rejection of roll-stock materials be made on a much more pragmatic basis than lots of individual packages such as cans, bottles, and cartons. Each roll is both an individual specimen (with respect to crushing or end damage) and a lot of repeats (with respect to defects that can affect a limited number of repeats within the roll, such as skips, tears, bad printing, splices). A shipment of rolls can be treated as a lot only in relation to inspection for roll defects; the rest remains a mystery until the rolls are individually run in production. In the examination of other materials, a quality audit is possible, if not always easy; in roll stock an audit is not practicable at all—this is the reason for placing extra importance on supplier quality control.

PAPER, BOARD, AND FLEXIBLE PACKAGE SPECS

SPECIFICATION FOR: Polyethylene Bag, 2-Pound Bread **NUMBER:** 72160
PROPERTY OF THE XYZ COMPANY   **EFF. DATE:** March 1, 1982

I. SCOPE: This specification describes a plastic bag which is intended to contain one 2-pound loaf of XYZ Sliced White Bread, wrapped in waxed paper, per Spec No. 72140. The corresponding product specification is No. 72100.

II. CONSTRUCTION
   A. The bag shall be made of low-density food-grade polyethylene film, with maximum clarity, so as to make the printing on the waxed paper wrap visible with no fogging. Film gauge shall be 0.002 ± 0.0005 inch.
   B. Overall dimensions shall be 20 1/2 inches long by 9 1/4 inches wide, with a 2-inch gusset at the bottom and a 1 1/2 inch header at the top. See Dwg. #25,* which is part of this specification. Gussets shall be sealed at the sides.
   C. Side seals shall be made with hot-wire cutters, and the seals shall be uninterrupted. Bag weight shall be 6.5 ± 1.0 grams.

III. PERFORMANCE
   A. Bags shall contain 2-pound loaves without rupture when subjected to six 30-inch drops, one on each side and end, when closed with a plastic/wire tie.
   B. In an air-inflation test, bags shall not rupture at less than 2.0 psig, and at higher pressures, the hot-wire side seams shall rupture at pressures no lower than the flat film.
   C. When used to hold 2 pounds of fresh bread without waxed-paper wrapper, the film bag shall limit weight loss at 70°F and 30% r.h. to note more than 2.0 grams per day over a 3-day period. In making the test, the bag shall be closed with a plastic/wire tie.

IV. MANNER OF DELIVERY: Bags shall be supplied in packs of 500 mounted on metal yokes. The yokes shall span 5 1/4 inches of the bag width at the header. On the packing line, bags shall peel downward from

---

*See Figure 18.

the headers with a force not exceeding 4 pounds. Prongs of the yoke shall pass through 1/2-inch holes in the bag headers. The holes shall be cleanly punched, with none of the punched-out circles falling into the bags.

Each case delivered shall contain four packs of 500 bags, each pack held together by the yoke at the header end and a paper band below the middle.

Cases shall be marked with the supplier's name, lot number, a plant location if more than one, and this specification number.

V. CLASSIFICATION OF DEFECTS
   A. Class A: defects which prevent the bag from being used or which frustrate its purpose of carrying and protecting the product:
      1. insecure mounting on yoke, which allows bag to fall on the floor. Bags which are contaminated by contact with the floor shall be discarded.
      2. holes, tears, or gaps in side seals
      3. odor, as from excessive extrusion temperature
      4. contamination, as with dirt, machine grease, or other deleterious foreign matter
   B. Class B: major defects which impair function or create a hazard of borderline performance:
      1. bag weight below minimum, or gauge of film below minimum
      2. film haze or fog, so as to impair readability of copy on the wax overwrap
      3. stains on the outside of the bag
      4. film circles cut from header on the inside of the bag
      5. low count on yoke
      6. dimensions outside of limits shown in Print #25 (Figure 18)
   C. Class C: minor defects of appearance or function:
      1. film torn or rough cut at the header
      2. stacks of bags untidy on yoke as delivered, so as to require straightening by hand
      3. scratches on film surface
      4. wrinkled bags

VI. ACCEPTABLE QUALITY LEVELS
    A. Class A: 0.1%

PAPER, BOARD, AND FLEXIBLE PACKAGE SPECS

    B. Class B: 1.0%

    C. Class C: 4.0%

VII. INSPECTION: The supplier is expected to conduct adequate inspection and exercise sufficient quality control over his product to assure compliance with the requirements above. He will also be assumed responsible for defects caused by his carrier who delivers shipments of bags to the XYZ Company.

Should it appear to XYZ personnel that any lot of bags may exceed the AQL limits, the Plant Quality Manager may conduct an audit. For this purpose he shall determine the lot size from the supplier's bill of lading and lot identification, and he shall sample the lot so defined in accordance with MIL-STD-105D. Lacking this information, he shall treat the truckload in question as one lot and will sample accordingly.

If inspection shows that the lot does exceed the allowable defect level, in one or more classes, the lot will be rejected to the supplier for his sorting or replacement.

Comments on the Bag Specification

Plastic bags are usually made at high speed on a single line of equipment; in this respect the quality-control aspects of the situation resemble the fabrication of can bodies. If something goes out of control, it is likely that a cluster of defectives will be produced all at one point in the lot. Single defectives would most likely occur when a roll of film feeding the bag-maker runs out, and a new roll is spliced in. There could also be some side-seam defects under startup conditions, before the hot-wire cutter/seamer reaches proper temperature, or if it accumulates a buildup of plastic.

The point to be observed here is that it pays to understand the manufacturing process and equipment in postulating possible causes of defects. This knowledge permits greater facility in exercising good judgment in the design of a sampling plan.

II. COEXTRUSION SPECIFICATIONS

Coextrusion is a process that assembles multiple layers of two or more molten plastic materials in a single operation to form a composite film or sheet. After some years of early development in several directions, the commercial process has evolved generally into single flat die extrusion with feedblock layering (Figure 19). The range of products used in packaging extends from two to seven layers composed of two to four polymer species, in gauges which commonly span 27 to 420 microns (0.001 to 0.150 inches), although a film of 100 layers can be made as thin as 14 microns (0.0005 inch) and appliance parts are thermoformed from sheet extruded at 560 microns (0.20 inch).

The advantages of coextrusion are that, with one extruder for each polymer species, one can

> combine any polymers with similar melt viscosities
> use tie layers (adhesive polymers) to combine otherwise incompatible polymers
> alter layer positioning and thickness by simple adjustment
> use one facility to make an infinite number of combinations and specifications for particular end uses

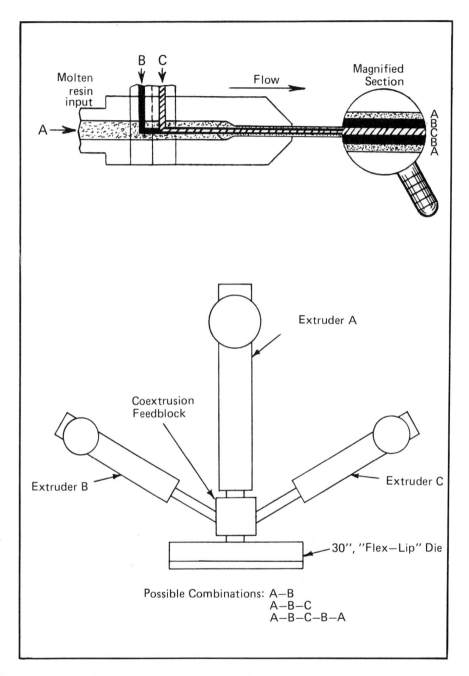

Figure 19 Schematic of coextrusion hardware.

Figure 19(b) illustrates the plan view of a facility with three extruders (three polymer species). Extruders B and C can be and frequently are mounted parallel to and above Extruder A, to conserve space.

The economies of coextrusion relative to lamination become obvious when one looks at a couple of examples (Figure 20). The one labeled 20(a) (polystyrene/PVDC/polyethylene) requires four extruders, including one for the tie layers of adhesive on each side of the central PVDC layer. The structure labeled 20(b) needs only three extruders. By coextrusion they are assembled in a single pass; by lamination it requires that the outer layers be separately extruded and primed for adhesion to the PVDC. Then, the PVDC must be applied as a light solvent coating and dried on one of the primed outer layers, or a minimum-thickness film grade of PVDC must be fed into the final lamination to form a composite structure. All this requires different extruders for the polyethylene and the polystyrene, a coater and dryer, and a laminator. There is little flexibility as to the amount of PVDC which can be applied in the laminated structure, while in the coextrusion the gauge of all components as well as the total is infinitely variable.

To match the supplier's sophistication in his capability to deliver highly specialized coextrusions, the buyer needs the knowledge to specify the best structure to suit his product and cost target, and the ability to monitor the conformance of deliveries to

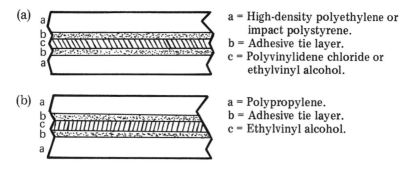

Figure 20 Illustrative coextruded film or sheet sections. (a) Polystyrene/PVDC/polyethylene. A three-resin coextrusion for gas and water vapor barrier. Good for stiffness; low heat resistance. (b) Polyolefin/PVDC/polyolefin. A high-heat resistant coextrusion with gas and water vapor barrier.

his specification. For example, the shelf life of his product may depend on a coextrusion in which a central layer of EVAL (ethylene vinyl alcohol) 3 microns in thickness provides the gas barrier, and the EVAL in turn must be protected from moisture by a polyolefin layer on each side not less than 50 microns in thickness. The buyer must be assured that the thin layer of EVAL is present and in the correct position relative to the other layers.

There are two laboratory procedures by which this assurance can be established: microscopic measurement of a cross-section of the coextrusion, and infrared absorption spectroscopy. The former uses less expensive hardware, costing less than $10,000, consisting of a microtome to provide a clean edge surface of film or sheet to be examined, and a microscope with micrometer scale to view the specimen, with suitable lighting.

The equipment for infrared spectroscopy costs four to five times as much as that for microscopic evaluation: the output is a graph of transmission versus wavelength, which is in effect a "fingerprint" of the material under investigation. A comparison of this graph or the absorption at key wavelengths for each ingredient of the coextrusion against a standard for the spec reveals whether each layer is present, and its thickness relative to standard. Figures 21 through 24 illustrate the "fingerprints" of common coextrusion ingredients. By comparing the graphs, one can grasp a few criteria for evaluating film and sheet:

- Thickness of polyethylene can be determined by noting absorption at 14 microns wavelength (Figure 21).
- Polyethylene and polypropylene can be distinguished by the absorption at 7.5 and 10 microns (Figure 22).
- PVDC is detectable by its strong absorption at 9.5 microns (Figure 23).
- Polystyrene is detectable by strong absorption at 13.5 and 14.5 microns (Figure 24).

In like manner, every other polymer will exhibit its own unique "fingerprint," and so will additives such as impact modifiers, plasticizers, slip agents, etc. Of the two test methods, the infrared spectroscopy shows formulation changes readily, while the microscopic test shows only numbers of layers and their respective thicknesses. The infrared scan will not show whether the several layers are in their proper positions in the coextrusion.

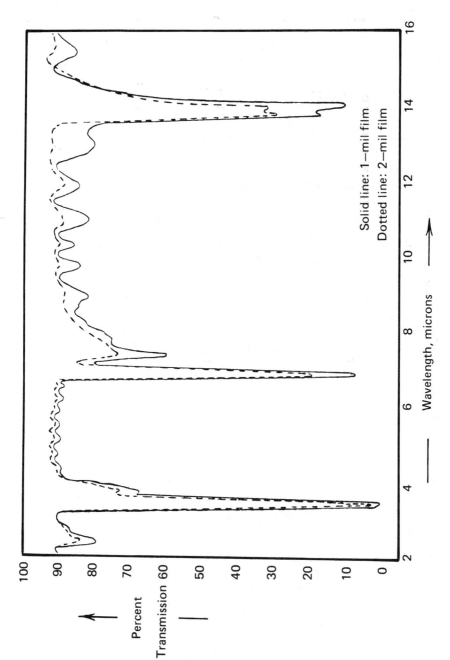

Figure 21 Infrared spectra, 25 and 48 micron polyethylene films. (Courtesy of Professor James Goff, Michigan State University.)

PAPER, BOARD, AND FLEXIBLE PACKAGE SPECS

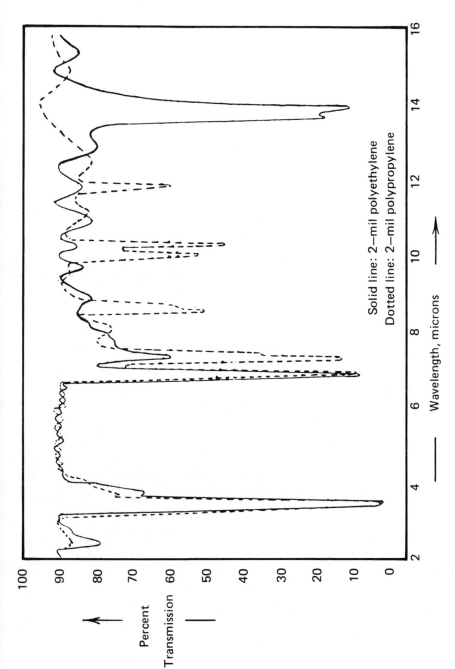

Figure 22 Infrared spectra, polyethylene and polypropylene films. (Courtesy of Professor James Goff, Michigan State University.)

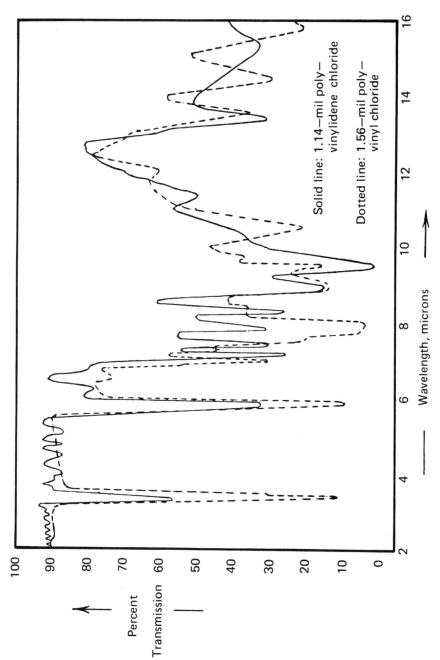

Figure 23 Infrared spectrum, PVDC (with PVC). (Courtesy of Professor James Goff, Michigan State University.)

PAPER, BOARD, AND FLEXIBLE PACKAGE SPECS 171

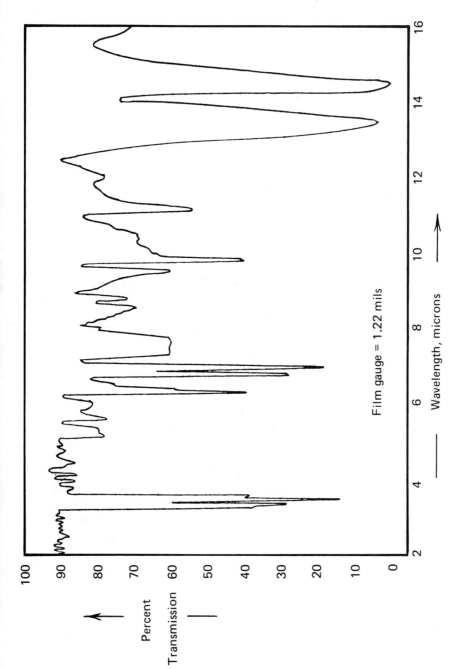

Figure 24 Infrared spectrum, polystyrene film. (Courtesy of Professor James Goff, Michigan State University.)

Overall, some new skills are involved for the buyer who wishes to make quality audits on coextruded packaging materials; in the long run, the benefits and cost advantages of coextruded materials far outweigh the need to acquire the skills and laboratory equipment if the end use is critical in terms of product quality delivered.

The most convenient way to specify coextrusions is by "percent of total thickness per component." A simple example would be a spec for thermoforming sheet to be made into a tray. The requirements, let us say, are for both water-vapor and oxygen barrier, and heat stability up to boiling, or 212°F. The structure would be that of Figure 20(b), and the total sheet thickness would be 0.025 inch. The spec, then, could be:

Layer	Percent	Thickness (inches)
Polypropylene	40	0.010
Adhesive	4	0.001
PVDC	12	0.003
Adhesive	4	0.001
Polypropylene	40	0.010
	100	0.025

Although 3 mils of PVDC may sound like a lot, the barrier will depend on its gauge at the thinnest part of the final thermoformed tray. All layers will thin proportionally in drawing; the thinnest spot may be 20% of the original sheet gauge. In that area, the PVDC will be 0.00067 inch, and the total thickness will be 0.005; these numbers represent good barrier properties.

The proportionality can be controlled down to thin films, but the percentages need to be altered to achieve the required barrier. Thus, a 3-mil film could be structured as:

Layer	Percent	Thickness (inches)
Polypropylene	33	0.0010
Adhesive	7	0.0002
PVDC	20	0.0006
Adhesive	7	0.0002
Polypropylene	33	0.0010
	100	0.0030

PAPER, BOARD, AND FLEXIBLE PACKAGE SPECS

The PVDC component of the film is about the same thickness as in the thinnest spot of the formed tray, while the total gauge of polypropylene in the film is half that of the total in the thinnest areas of the tray: 2 mils versus 4 mils, respectively.

In the quality inspection of coextruded sheet and film, defects can be classified as follows:

Class A. Critical
   1. Contamination
   2. Missing layer in the structure
   3. Tears, holes, or gaps
Class B. Major
   1. Total gauge below minimum
   2. Any layer less than 80% of nominal gauge
Class C. Minor
   1. Scratches on film surface

The above classification gives the most important specifics for coextrusions. The list of defects can be reinforced with those common to all flexible materials, such as the list applied to the 2-Pound Bread Wrapper, Spec #72140, on prior pages of this chapter.

III. METALLIZED SPECIFICATIONS

Paper, paperboard, and plastic films have for the last 5 years been applied in large commercial volume with condensed aluminum vapor coatings for packaging end uses. The metal vapor is generated by boiling aluminum metal in a high-vacuum chamber and allowing it to deposit on the cool surface of smooth plastic or paper as it is reeled across the evaporating surface.

Applications on cartons are mostly for the appearance of mirrorlike silvery reflectance; as such, specific end uses are in the cosmetics and toiletries market segments. The effect is obtained by laminating a transparent metallized film—usually polyester—to the outer surface of the carton board, followed by surface printing. The combination is readily embossed as well.

Applications on paper are mostly for labels which would otherwise be foil-laminated. The advantages of metallizing over foil are twofold: ease of penetration by detergent solutions in washing labels off of returnable bottles, and much less problem with labels curling before application when shipped or stored in conditions of changing humidity.

Film applications represent the largest volume market segment for all metallizing, i.e., flexible packaging end uses. In this environment, the metallizing process imparts a true barrier function increment to the base film; if film is required in the spec for other reasons, such as strength, heat sealing, and printing, then metallizing provides the barrier enhancement at a lower cost than foil laminating.

To specify metallizing for packaging, the specialist must be informed about its applicability to given substrates, the extent of barrier enhancement that is feasible (when needed), and its durability through converting, packing, and distribution. For end uses where mirrorlike appearance is the primary need, a deposition of 100 Angstrom units in thickness (10^{-8} meter) is adequate, whereas the optimum for water-vapor and gas barrier is about 300 Angstrom units. The latter is 4/1000 the thickness of 0.00027-inch (7-micron) aluminum foil; it is apparent that one is dealing with very thin coatings of aluminum.

Thus, two variables and one attribute are critical to the barrier performance of metallized films: the variables are thickness of deposition and adhesion to the substrate; the attribute is freedom from damage due to abrasion or scuffing. Thickness of deposition is too low for measurement by common mechanical gauges; research has led to the development of a light-transmission tests as a parameter to amount of metal deposited. Figure 25 shows a relationship between barrier property and optical density, as published by the (American) Association of Industrial Metallizers, Coaters and Laminators (AIMCAL). Off-the-shelf instruments for measuring optical density are available as quality-control testers.

Adhesion of a metallized coating to its substrate can be measured in the same way as lamination strength of a multilayer flexible material. In its crudest form, and for quick quality assessment, a pressure-sensitive tape is pressed onto a metallized surface and peeled away. If the metallizing is pulled away with the tape, adhesion is weak.

To maximize the barrier qualities, metallized film is coated with a sealing lacquer on the metal side or laminated to a heat-

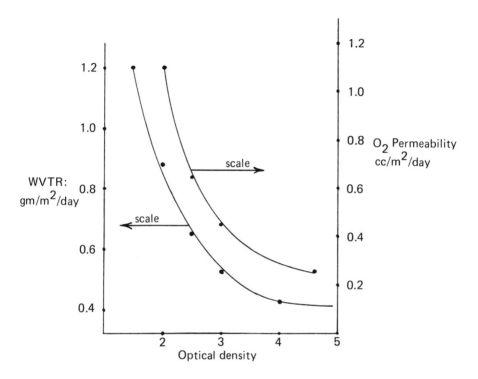

Figure 25 Water vapor and oxygen barrier, metallized polyester film. (Courtesy of Camvac International, Inc.)

sealing film with as little handling as possible after the metallizing step. It is important that this be done soon after metallizing, also, since the thin layer of deposited aluminum will oxidize if stored in humid conditions. Obviously, the handling is minimized to prevent scuffing the metal layer before it is sealed in by coating or lamination.

Metallized film has developed a large share of the market for the flexible packaging of snacks and confections, especially those products which contain fats which can oxidize and lose freshness unless gas-packed. Metallized film brings two advantages to such packaging: maintenance of good gas barrier to block oxygen entry and light barrier to block ultraviolet stimulation of fat oxidation.

For granular and powdered product packaging, metallized film has the further benefit of grounding static electricity, which minimizes seal contamination and permits higher-speed packing operations. This property is so prominent that metallized-film pouches are used for packing almost all computer microchips with printed circuits, to prevent damage by stray currents and static charges.

As an example, the spec for a pouch pack of ground coffee to be used in restaurants could state:

> MATERIAL: The pouch shall be composed of 50-gauge clear polyester film, metallized to an optical density of 3.0, and laminated on the metal side to 2.0 mils of LDPE blown film with 0.5 mil ionomer film extrusion. The resulting stock shall be surface-printed on the polyester side. Oxygen permeability of the finished stock shall be 0.5 cubic centimeter/square meter/day maximum and water vapor permeability shall be 0.6 gram/square meter/day maximum.
>
> Adhesion of the metal layer can be tested by a Gelbo flex test, which will determine durability of the metal/polyester bond at the interface with the ionomer layer.

PAPER, BOARD, AND FLEXIBLE PACKAGE SPECS

SPECIFICATION FOR: 12-Ounce Detergent Carton   NUMBER: 31270
PROPERTY OF THE XYZ COMPANY   DATE: August 1, 1982

I. SCOPE: This specification describes a fiberboard carton to contain 12 ounces of XYZ Powdered Dishwashing Detergent. The product is made to Spec No. 3000, dated March 12, 1980.

II. CONSTRUCTION
   A. The carton shall be made of 0.024-inch white machine clay-coated chipboard, medium density.
   B. Blank shall be dimensioned in accordance with Print No. 31207,* of above date, which is part of this spec. Carton shall have full flaps, seal ends, with Van Buren ears to prevent sifting, and a perforation pattern at the top of one side panel for easy opening and pouring.
   C. 1000 cartons shall weigh 143 ± 7 pounds at standard conditions.
   D. Manufacturer's joint shall be glued, 5/8 inch wide.
   E. Printing shall be four colors plus scuff-resistant varnish, in accordance with mechanical art specified separately. Grain direction of the paperboard shall be horizontal to the printing.

III. PERFORMANCE
   A. Cartons shall operate on Brand Z cartoning machines at 200 per minute.
   B. Break-open force of k. d. cartons shall not exceed 700 grams, and spring-back shall not exceed 300 grams.
   C. Filled and sealed cartons shall not allow product sifting when exposed to 1 hour of vibration on an L.A.B. vibrator in cases at 1 g acceleration.

IV. PACKING FOR DELIVERY: Cartons shall be stacked with uniform facing in layers of 500 in corrugated boxes. Each box shall contain six stacks, separated by chipboard pads. Boxes shall be stenciled with supplier's name and plant number, this spec number, and lot number, with date of manufacture.

---

*See Figure 26.

V. CLASSIFICATION OF DEFECTS
  A. Class A: faults which prevent the cartons from containing the detergent, or from identifying it.
    1. Excessive opening or springback force
    2. dimensions outside of tolerances shown on Print No. 31270
    3. tears, holes, or scuffs that mutilate or make illegible any of the copy
    4. missing color or colors
    5. misregistration of colors such as to make any copy illegible
    6. misregistration of scores such as to cause production difficulty in setup, filling, or sealing
  B. Class B: faults which cause marginal function or low-quality appearance:
    1. smudged printing or scuffed surface showing board through the clay coating
    2. uneven scores, making carton setup difficult on the packing line, causing loss of packing efficiency
    3. inadequate perforation for glue penetration on seal flaps or Van Buren ears
    4. inadequate perforation for ease-opening feature on carton side
  C. Class C: minor faults affecting appearance only:
    1. rough printed surface or low gloss from varnish
    2. slight off-color match

VI. ALLOWABLE DEFECT LEVELS
  A. Class A: 0.4%
  B. Class B: 1.0%
  C. Class C: 2.5%

VII. INSPECTION: The supplier is expected to conduct quality control and inspection sufficient to assure compliance with this specification. Should it become apparent because of poor carton performance that a lot (a truckload, about 500,000 cartons) may not comply, the Plant Quality Manager may call for an audit. If sampling done in accordance with MIL-STD-105D and inspection made in accordance with this spec show defect levels over the limits in VI, above, the lot shall be rejected to the supplier or set aside for his inspection and sorting.

PAPER, BOARD, AND FLEXIBLE PACKAGE SPECS   179

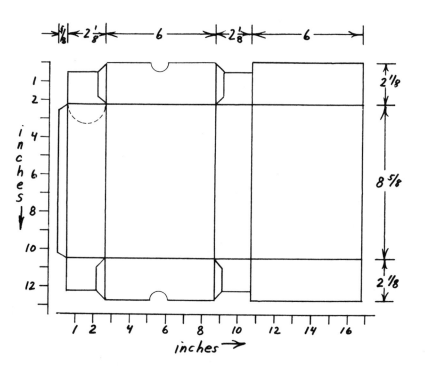

Figure 26 Sample specification: Keyline, 12-ounce detergent carton, printed side.

IV. NOTES ON QUALITY ASPECTS OF CORRUGATED SHIPPERS

All the unit packages discussed in this and earlier chapters would be distributed in corrugated boxes, with very little likelihood of exceptions. The corrugated box is, in fact, the most widely used shipper of any kind, in industrialized countries. No matter what the final product contained, the shipper must provide resistance to impact, abrasion, puncture, and dirt penetration, must identify the product and shipper or the consignee, and must provide stacking strength if the unit packages do not; all at minimum cost.

One shipper spec was described in Chapter 5; with the information included at that point and in this section, it will be possible for the packaging specialist to identify the quality considerations for corrugated shippers to protect almost any product in its unit packages.

First, let us classify defects in corrugated boxes, as we have done it for several other kinds of packages:

A. Class A (or critical): defects which frustrate the function of the box and prevent it from safely storing and distributing the unit packages of product it contains:

1. incomplete liner adhesion
2. dimensions outside of tolerances for the unit packages it must hold (limits ± 1/16 inch)
3. weight below minimum
4. basis weight of a liner or medium below minimum
5. loose manufacturer's joint

B. Class B (or major): defects which reduce box function to a marginal level, such that it is likely to fail under stress, although it may perform adequately under ordinary conditions of storage and distribution.

1. incompletely glued manufacturer's joint or incomplete tape joint
2. deep slots, running down into the edges between ends and sides
3. outer flaps not meeting by gap exceeding 3/16 inch
4. bad scores
5. excessively high or low moisture content in board (below 5% or above 30%)
6. no nonskid treatment when specified

PAPER, BOARD, AND FLEXIBLE PACKAGE SPECS 181

C. Class C (or minor): defects which reduce the quality of box appearance, but not necessarily its function:

1. rough-cut slots and/or flaps
2. washboard appearance, making for poor printing
3. visible specks of nonpaper content in the outer liner, such as tar, asphalt, chaff, or dirt particles
4. bad printing for any other reason

Some of these defects are likely to occur in clusters, and others will usually appear randomly. Clustered defects will arise from a machine maladjustment, such that a lot or part of a lot is produced out of spec, for example. Random defects will be made by fleeting conditions in box-making, such as a tear in a roll of linerboard or a brief interruption of glue feed at a spot in the corrugator. As we saw in Chapter 4, random defects are less likely to be found in a standard sampling plan than are systematic defects; first, because the former are probably less numerous. If, on sampling a lot received, a defective is found on which there is doubt whether it be systematic, the best step is to look at the two boxes on either side of the defective found. If the latter also have the same defect, it is probably systematic, or the result of a temporary process fault that was not noticed by the supplier. To be able to pick up specimens in a bundle adjacent to a first sample implies, of course, that the location of the first sample is recorded, so that one knows where to go for the second sampling.

From the list of defectives above in Class A, B, and C, we can further classify them as to probably random or probably systematic:

Probably random
    Incomplete liner adhesion—usually arises from interruption of glue feed
    Loose or incomplete manufacturer's joint—from interruption of glue feed
    Bad scores—from light weight in liner board, more common on corrugated of 150 test and lower
    Bad printing—from worn flexo dies or low ink feed, or washboard corrugated
Probably systematic and clustered
    Dimensions outside of limits—wrong cutting die

Low weight—wrong weight roll of kraft for one or both liners
Basis weight of liner too low—same as above
Short tape—wrong setting on tape feed
Deep slots—wrong setting on slotter knives
Outer flaps not meeting—top and bottom scores too far apart—box may be too high
Moisture content out of control—poor storage conditions for finished boxes
No nonskid when specified—interruption of spray feed

Box orders are usually set up for one printer-slotter, and an entire lot will be produced on the one machine. If there is an error in the selection of the corrugated stock, the setup of the die, or the adjustment of the slotting knives, every box will be off in weight or dimensions or slot depth until the error is corrected. On the other hand, if everything is correct at the start of the lot, it will very likely be correct throughout in those criteria. Attention is needed during the continued fabrication of the lot to see that the glue or tape feed for the manufacturer's joint is in control and that the nonskid spray, if required, is not interrupted. The same applies to the ink flow for printing.

Every packaging specialist who writes specifications should visit at least one plant which fabricates the package or component types he is specifying, because an acquaintance with the supplier's process is necessary to get a "feel" for the kinds of material and process variables which can occur and cause quality defects in the materials he receives. Such knowledge can be most useful for designing a sampling procedure, or when a quality audit seems desirable on a lot of material that is suspected to contain more than the allowable defective levels.

9

Unitized Constructions for Distribution

Manufacturers place total dependence on approved shipping containers to get their products safely from packing plant to point of use. For their part, carriers back up approved shipping containers with their insurance. Nevertheless, for economic reasons the individual corrugated box or kraft bundle or multiwall bag is not necessarily the final step in packing goods for distribution.

Many products are indeed sent to their destinations in single bundles, tray cases, corrugated boxes and bags, hand-loaded into trucks, railcars, and even cargo planes. The process is called "rock piling" or "dead loading." Starting with a layer on the car or truck floor, the filled shippers are piled as high as the roof of the conveyance permits, or as high as the maximum weight load permits. The boxes or bags are stacked snugly together to prevent load shifting and damage in transporting. This procedure is less than satisfactory for its high labor cost at both ends of the trip: each of a couple of thousand bags or boxes must be individually carried into the conveyance and stacked, and the consignee must pay as much for removal. Depending on the circumstances, there may be two more drawbacks: if the shipment is less than full truckload or carload (LTL or LCL), the control over damage resistance is lost, and if the product is low in density, the loaders will have to build the stacks to the roof, possibly crushing the lower layers of containers while stepping on them.

The shortcomings of the dead loading system led to the development of the "palletizing" system, which had its beginnings in World War II and its greatest expansion during the 1950s and 1960s. The essence of palletizing is to stack shipping containers about 3 to 5 feet high on rigid platforms of fixed size, moving the loaded pallets into and out of trucks and railcars with fork trucks. If, on the average, each pallet can carry 50 individual shipping containers, the labor of carloading is cut by 98% versus the labor for dead loading. This assumes that the stacking of bags or boxes onto the pallets can be done off the end of the packing line with no more labor than would be used to remove product without pallets.

The pallet with shipping containers stacked on it is one form of "unit load." The components of a specification for unitized pallet load are:

1. The pallet itself—dimensions, material of construction, details of assembly, and truck access features
2. The pallet pattern—the manner in which and how many individual shipping containers are stacked on each unit load
3. Auxiliary means, if any, for stabilizing or holding the unit load snugly together; for instance, strapping

Let us look at each of these elements in some detail.

I. PALLETS

A pallet can be made of wood, corrugated and honeycomb paperboard, plastic, reinforced plastic, or metal. The choice is based on service conditions, such as weight of load, climatic environment, durability requirement, local availability, and costs. The most common of the hundreds of specifications are wood pallets, and the most common of these is the grocery industry pallet, a platform 48 × 40 inches, accessible to fork lift entry on all four sides. (See Figure 27.) There are 13 grade subspecifications covering

UNITIZED CONSTRUCTIONS FOR DISTRIBUTION

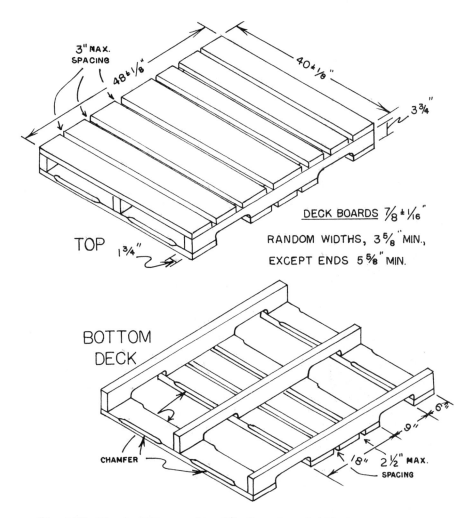

Figure 27 Wood pallet general specification, grocery industry.

variables such as number and type of nails, chamfer on the deckboards, and type of wood, all issued and administered by the National Wooden Pallet and Container Association, 1619 Massachusetts Avenue, NW, Washington, DC 20036.

II. PALLET PATTERNS

For any given dimensions of shipping container, there should be one best arrangement for laying units on a pallet so as to cover the greatest possible area, while minimizing overhang or underhang on the four sides. The best pattern is first established for one tier, or layer; the second tier should, if possible, be the inverse or mirror image of the first, to provide an interlocking effect, as bricks are laid in building a wall, that gives the unit load resistance to loosening with vibration and shock. The third tier should then be the same as the first, the fourth the same as the second, and so on. Figure 28A and B illustrates the principle.

To accomodate the almost infinite variety of individual shipping-container sizes, there have been devised almost 150 interlocking patterns for the common 48 X 40 inch pallet. There is, of course, no assurance that any given new case size can be perfectly fitted to even one of the interlocking patterns. The packaging professional who specifies the package must, if his company ships on pallets, identify the best pattern available and note it on the shipping container spec, so that the shipping department need not flounder when a new product spec starts coming off the packing lines.

Failure to identify a pallet pattern can produce chaos in physical inventory control if operators on different lines or in different plants use pallet patterns of their choice, the number of units per pallet will vary with time and location; accurate counts for inventory and shipment cannot be made by assuming a fixed number of units per pallet and simply counting pallets. The entire cost incentive for palletizing would be negated if each pallet load had to be disassembled for counting units.

III. AUXILIARY STABILIZING

There will be some shipping containers that cannot be palletized with any interlocking pattern; the most common examples are cases with square cross-sections. These must be "column stacked," that is, simply piled up in side-by-side stacks that have no interaction. In these situations some additional step must be taken to prevent the pallet loads from toppling in handling and shipping. Otherwise, simply driving a column-stacked load across a rough

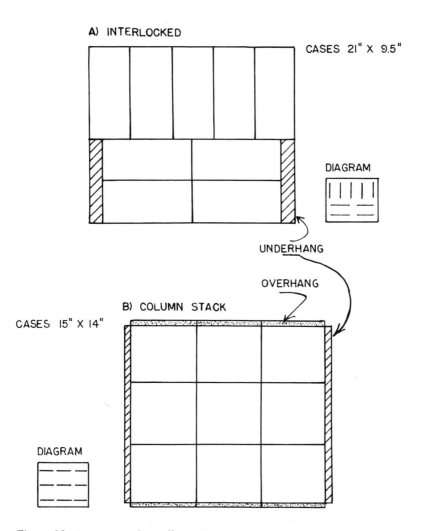

Figure 28 Representative pallet patterns.

warehouse floor on a fork truck is enough to scatter its units along the way. Several degrees of sophistication are available for stabilizing column-stacked pallet loads. The first step, of course, is to establish the best pallet pattern for space utilization for minimum overhang or underhang. Almost 50 such patterns have been identified for the grocery industry. Beyond that, investment for stabilizing means can range from zero to $100,000 per line for (in order of increasing capital):

1. Tying string around the girth of the top tier on each pallet load
2. Running strips of masking tape over the top and down the sides
3. Laying a chipboard or kraft sheet between successive tiers
4. Same as #3, adding a light striping of low-tensile glue on the top surface of each interleaving sheet before mounting the next tier of shipping containers
5. Covering part to all of the pallet load with a paperboard or corrugated cap or shroud, stapled or taped along the edges
6. Fastening the load to the pallet with loops of plastic or steel strapping
7. Stretch or shrink wrapping the pallet load with plastic film—from a hand operation with heat-shrink gun to total automation with a conveyorized heat-shrink tunnel
8. Any combination of the above

Choice of the optimum spec can be based on the nature and value of the product, the cost implications of damage, the storage and distribution environment, and, most important, the safety of the personnel involved with handling, warehousing, and shipment. Whether interlocking or column-stacked, loads that may topple on someone when stored three pallets high in a warehouse, or when a railcar door is opened at point of receipt, cannot be tolerated.

With the information above, it is possible to draft a palletizing spec, for example using the 20-ounce soup pack of Chapter 6. The case *outside* dimensions are (page 120) about 14 × 10 1/4 × 5 inches. They can be laid on a 48 × 40 pallet in a 14-case interlocking pattern, thus:

UNITIZED CONSTRUCTIONS FOR DISTRIBUTION 189

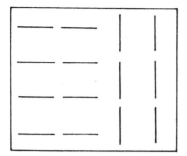

The cases will overhang the 40-inch sides by about 2 inches and the 48-inch sides by about 1 inch. The interlock pattern may require no auxiliary stabilizing, but the packaging engineer must make a choice. His knowledge of the annual volume for this product, history of damage, shipping-department labor costs, space available, customer preferences, and materials costs will lead him to a decision.

Let us say that after consultation with the department and the purchasing agent he runs some tests and decides in favor of a corrugated cap. The palletizing spec then looks like this:

SPECIFICATION FOR: Unit Load, 20-Ounce Soup   **NUMBER:** 520813
PROPERTY OF THE XYZ COMPANY   **EFF. DATE:** October 1, 1984

I. SCOPE: This specification covers materials and procedures for palletizing cases of Spec No. 520812, dated October 1, 1984, containing 12/20-ounce canned soup.

II. MATERIALS
   A. 48 X 40 Grocery Industry 4 Way Pallet
   B. Corrugated cap, nontest, C-flute, 42 X 49 inches i.d., with 7-inch sides, stapled corners

III. PROCEDURE: Place cases right side up on the pallet according to Pattern #100* (14 cases per tier). Interlock six tiers high (30 inches). Set a corrugated cap over the top so as to completely cover the uppermost tier and part of the next lower tier.

---

*It is helpful to the shipping department if the pallet pattern is printed along with other necessary copy, on the top of each case, in the form of a small diagram.

UNITIZED CONSTRUCTIONS FOR DISTRIBUTION

The importance of overhang and underhang should be discussed; they have an immediate effect on ease of loading and unloading trucks and railcars, and a decisive bearing on damage in transit. Figure 29 shows the plan view of a nominal 50-foot railcar and a 40-foot trailer. On the plan are superimposed a layout of 48 × 40 pallets as they are intended to be positioned, in a full carload and truckload, of course. When weight limits permit, the pallets are stacked two high in either type of conveyance, provided that individual unit loads, including the pallet, do not exceed 50 inches in height.

In a railcar, pallets are loaded with 48-inch sides facing the ends of the car, and 40-inch sides the car sidewalls and centerline. The 50-foot car will carry a total of 56 pallets.

The road trailer, which is limited by law as to width, can hold 40 pallets in a nominal 40-foot size. One row is loaded in 11 pallet positions along one side, with the 48-inch sides facing the ends of the trailer. The other side is loaded in nine positions, with a 40-inch sides facing the trailer ends.

The problem of too much overhang is obvious—the unit loads will be too big to maneuver into position. Even worse, if they are loaded snugly, the loads will settle a little as a result of vibration and vertical shocks in transit, wedging themselves in so tightly that they cannot be removed intact.

Excessive underhang (more than bout 2 inches on a side) will allow too much movement of loads without auxiliary stabilizing, even though they are built with an interlocking pallet pattern. Consider that 2 inches of underhang on a side multiplies to 56 inches of empty space along the length of 14 pallets in a railcar. A sudden stop will allow cases to shift one whole pallet length to pile up in the door space and one end of the car. Bulkhead doors and dunnage cannot take up the space left by underhang. There is no alternative but to stabilize each pallet load by such means as noted above.

IV. DEVELOPMENTS IN UNIT LOADING

Pallets have done and are doing a very effective job of carrying millions of tons of merchandise per year, but their cost of ownership has tripled in the past 10 years, and they are not always easy

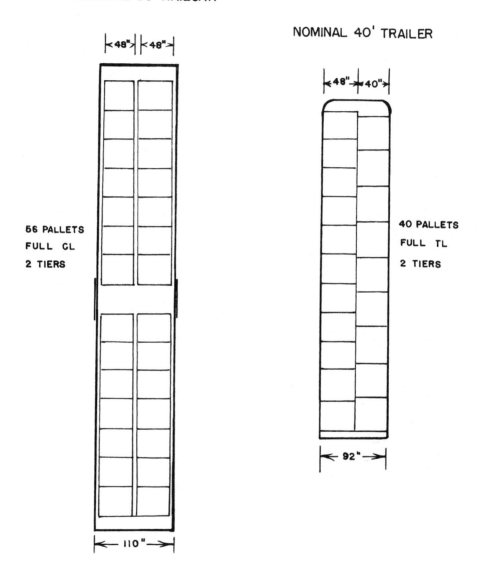

Figure 29 Railcar and trailer pallet positioning plan.

to recover once sent out. They are breakable if handled roughly, they occupy storage and transit space, and if not kept clean, they can carry insects and dirt from one plant to another.

The unit-loading ideal would be a one-way platform so cheap that it can be discarded after one trip, and flexibly sized to match without overhang or underhang the dimensions of a good loading pattern for whatever shipping containers it carries.

Such a low-cost platform, known as the slip sheet, has been commercially developed, but while it is making inroads toward taking over the work of rigid pallets, it cannot be handled by ordinary fork truck equipment, nor can it replace the pallet in rack storage without rack modification. Meanwhile, palletizing has undergone considerable refinement, from the pallet construction itself (e.g., nestable plastic pallets, which halve empty pallet storage space) to the improvement of unit load stability with shrink film and stretch film. The latter improvements have been adopted in turn to slip sheet unitizing to complete the picture.

Several current versions of commercial unit loads are shown in Figure 30. A brief description of each is necessary:

A. Flow-Through Shrink Wrap

A pallet is passed between two wide rolls of shrink film which are mounted on vertical axes, and which have been joined by a heat seal. The pallet passes as through a gate, pushing the film ahead of it and along its sides. The film is pulled in behind the pallet after its passage between the two rolls, by a pair of heat-sealing bars. The pallet is now loosely wrapped around all four sides in a film which extends downward over the pallet and upward over the top tier of cases. With exposure to heat for a minute or two—radiant, or tunnel, or hot-air blowers—the film shrinks snugly around the pallet. The shrink-wrapped unit load can be handled with a conventional fork truck.

B. Spiral-Wound Stretch Wrap

The palletized load is set on a turntable beside a single roll of stretchable film 18 to 24 inches wide, also mounted on a vertical axis. The film end is tucked between a couple of cases, and the turntable is started, causing the pallet to pull film off the roll onto

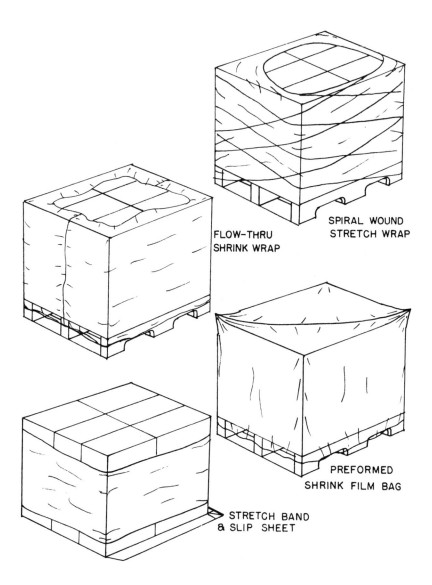

Figure 30 Types of film-wrapped unit loads.

itself. The axle holding the roll of film is powered and programmed to move up and down while the film is being pulled off, causing a top-to-bottom spiral wrap to form around the pallet. The film axle is also braked so that the film is stretched more than 15% as it is pulled onto the pallet load. The operator can apply as many wraps or plies as desired for the weight and distribution pattern of the product in question. Handling is accomplished by fork truck.

This system has a couple of advantages over the flow-through shrink wrap: film application facilities are somewhat cheaper, heat shrinking is unnecessary, and the space required for the operation is correspondingly less. The traverse of the narrower film can readily be adjusted to pallet loads of different heights, while the flow-through process requires changing film rolls to match width against pallet load height. Finally, if the individual shippers which make up the unit load consist of shrink-film tray packs, the pallet stretch wrap will not stick to the shrink film underneath. This has been a problem with pallet shrink wrap over tray shrink wrap, caused by the heating step.

C. Preformed Shrink-Film Bag

When numbers of pallet loads of the same height are to be unitized, and it is desired to minimize space and equipment costs, preformed bags of shrink film can be purchased. One or two men can apply the bags and shrink them with heating equipment of any sophistication; the packer is trading capital investment (for flow-through equipment) against a higher labor cost. The chief advantage of this system is that the palletized goods can be stored outdoors, since the top as well as the sides is covered. This assumes that the product is insensitive to heat or cold, of course.

D. Stretch Band and Slip Sheet

Figure 30 really shows two developments in combination and, unlike the other three, represents a unit load which cannot be handled with conventional fork trucks.

The shrink band is merely a modification of the spiral-wound stretch wrap, wherein the vertical traverse of the film roll is reduced so it does not cover the unit load over its full height. This represents the most economical use of stretch film, in that only enough is used to get around all the cases.

The major innovation is the replacement of the pallet with a heavy sheet of kraft board, 0.060 to 0.070 inch thick, with a lip a couple of inches wide protruding from two adjacent sides. The equipment that makes the slip sheet feasible is a modified truck with (1) a pincer which can grip the slip sheet lip along its entire length, (2) broad, polished wedge-shaped platens in place of fork tines, onto which the pincer can pull the unit load by its slip sheet, and (3) a vertical pusher plate with which the truck operator can discharge the unit load off the platens. The unit is called a "Push-Pull" or "Pull-Pack" truck.

Major manufacturers offer conversions from fork to pull-pack trucks, so that total obsoletion can be avoided.

The advantage of the slip sheet is its low cost—less than 10% of the outlay for a new wood pallet. Furthermore, the slip sheet can be supplied to the size of a "pallet" pattern, avoiding overhang and underhang entirely. In unobstructed warehouse space, as with cantilevered roof, this can lead to great economy through closer stacking. It should be noted, however, that for efficient loading of trucks and railcars, one cannot depart too far from the 48 X 40 dimension without running into excessive void spaces or tight fits. At the time of this writing, the major factor determining proliferation of the slip sheet is the dispersion of pull-pack equipment. A packer cannot ship on slip sheets if his customers cannot handle them at their end, but the incentives for slip sheets—no return, low space need, low cash outlay—are so strong that expansion of their use is quite certain. They are not rigid enough, of course, to act as self-supporting platforms, as do pallets, in rack storage. To adapt to slip sheets, the racks must first be equipped with shelving strong enough to bear the unit loads.

V. CLAMP TRUCKS

Where products and their shippers are strong enough to stand squeezing, or sidewise compression, without distortion or breakage, unit loads can be handled by clamp truck. This equipment consists of two flat vertical steel plates at the sides of the truck instead of forks at the bottom. The operator can grip and lift a unit load between the plates, adjusting the hydraulic plate pressure by a gauge, according to the weight of the load.

In such a system, a pallet is not required as a platform for handling, nor is a slip sheet, but the load must be compression-resistant. For adaptable products, the clamp-handled unit load can be the ultimate in materials economy.

VI. REGULATIONS

When individual shipping containers are composed into unit loads of any kind, they are usually less subject to damage in transit, and particularly so if they are reinforced by auxiliary stabilizers, such as film or caps or shrouds. This does not permit relaxation of the minimum specs for the shipping containers as prescribed by Uniform Freight Classification (rail) and National Motor Freight (truck) rules. The presumption of the carriers is that at some point the unit loads will be broken down into their component cases, still within the responsibility of a common carrier.

Future history may show that this is not a significant concern with carload and truckload shipments, which would open the door to some economies in shipping containers when used only in unit loads. Until evolution occurs in this direction, the supplier of shipping containers remains the key to compliance with regulations.

A supplier always has experts on the Uniform Freight Classification who can not only advise on the requirement, but also have the facilities to make and test samples at low cost and in short time—in the magnitude of a couple of hundred dollars and a couple of weeks, usually. While the Uniform Freight Classification states a *minimum* shipper spec, the customer's product may be constructed in such a way that the minimum is inadequate. To be sure, a common carrier's insurer will pay damages on products shipped in minimum-spec cases or bags, but damaged goods result in lost sales and loss of confidence in the product by its distributors and sellers.

Certainly the shipper supplier must be advised if there is to be anything unusual about the distribution of the product in question. In the absence of other information he will ordinarily design on the assumption of rail shipment. This is normally more than adequate for truck distribution, but not adequate for express or parcel post, which impose rougher handling on packages than either rail or truck.

VII. TESTING

The evaluation of damage resistance has been undergoing both intensive study and radical change within the last 10 years, and the cohesion of the whole picture is not yet complete. All that has thus far been noted on quality related to the individual package components, while damage resistance and the physical condition of the primary package at point of sale depend on the combination of the performances by the primary package, the shipping container, and the unit load.

For most of distribution history, that is, from the Industrial Revolution to World War II, most damage-resistance testing consisted of rolling filled shipping cases down a flight of stairs or dropping them from an arbitrary height. The broadscale use of simple drop- and impact-testers and vibrating tables blossomed in the late 1940s, while systematized protocols for applying them had been developed by the end of the 1950s. It is important to note that the current specifications for shipping-container materials, and the criterial for damage assessment, were jelled during those two decades; thus, carrier rules for acceptance to ship and acceptance of damage penalties were built around dead-load distribution.

Early in the 1960s, however, unitizing became general practice, mostly with pallets, when there was neither readily available equipment nor a set of procedures to test the adequacy of unit-load integrity and damage resistance of products in unit loads. Since one could not readily shove a unit load down a flight of stairs, nor did it seem appropriate, only two tests of an unofficial nature were used commonly:

- Static compression, by stacking unit loads up to five high in a warehouse
- Shock, by lifting one end of a unit load 6 or 8 inches with a fork truck and letting it fall

This being the case, the packaging engineer had to depend on the "shipping test" for realistic evaluation of unit load performance.

UNITIZED CONSTRUCTIONS FOR DISTRIBUTION

It soon became obvious that the "real world" is highly variable, such that shipping tests were required to be run in four conditions:

- Carload (CL) and less-than-carload (LCL) via rail
- Truckload (TL) and less-than-truckload (LCL) via truck

Within that fourfold environment, the forward end of a railcar gave the most severe horizontal shock forces, and the rear end of a truck (aft the wheels) the most severe vertical shocks. Thus was born the intensive use of ride recorders, to chart shock intensities in three dimensions as a means of estimating the abusiveness of a given trip.

Through the 1960s and early 1970s, launching the first production of a new product/packaging combination was a worrisome thing. Lab tests of single shippers could look good, and the first eight or nine carloads and truckloads could arrive at destinations damage-free, while the next would be heavily damaged in the top tiers of the upper unit loads of a railcar shipment. The knee-jerk reaction would be to heavy-up the shipping container board spec, which would not make any difference to performance; net, heavy costs were incurred for travel and inspection, and for "stronger" packaging, without confidence that the pack was safe.

By the mid-1970s, technology of testing equipment had been developed to the size and power of handling unit loads as well as shipping containers and, more important, had demonstrated the severity of damage that can be caused by vibrations in distribution at the natural frequency of a unit load or stack. This knowledge led to everyday application of vibration scanning over a range of frequencies to identify for a packer the natural frequency of the unit load. With this information, he could confidently modify the structure to snug or stabilize the unit load, usually without adding cost, and avoid magnifying the low-frequency vibrations generally experienced in transportation.

In 1981, the American Society for Testing Materials (ASTM),* after 8 years of study and trials, established a series of protocols for performance testing of finished goods, including unit loads, entirely within a laboratory. Any of 14 distribution cycles can be

---

*Committee D-10. See References for Chapter 9.

PERFORMANCE TESTING OF SHIPPING CONTAINERS

TABLE 1 Performance Test Sequence

Sequence	1	2	3	4	5	6
Distribution cycle 1	General schedule—undefined distribution system					
Test type	Handling	Vehicle stacking	Stacked vibration	Loose-load vibration	Rail switching	Handling
Element	A or B	D	E	F	H	A or B
Distribution cycle 2	Specially controlled environment, user specified					
Test type			USER SPECIFIED			
Element			Select from 11.1			
Distribution cycle 3	Single package environment, up to 100 lb (45.4 kg)					
Test type	Manual handling	Vehicle stacking	Loose-load vibration	Vehicle vibration	Manual handling	
Element	A	D	F	G	A	
Distribution cycle 4	Motor freight, single package over 100 lb (45.4 kg), or palletized					
Test type	Handling	Vehicle stacking	Loose-load vibration	Stacked vibration	Handling	
Element	A or B	D	F	E	A or B	
Distribution cycle 5	Motor freight, TL, not unitized or palletized					
Test type	Handling	Vehicle stacking	Stacked vibration	Loose-load vibration	Handling	
Element	A or B	D	E	F	A or B	
Distribution cycle 6	Motor freight, TL, or LTL—unitized or palletized					
Test type	Handling	Vehicle stacking	Stacked vibration	Handling	Warehouse stacking	
Element	B	D	E	B	C	
Distribution cycle 7	Rail only, bulk loaded					
Test type	Manual handling	Vehicle stacking	Stacked vibration	Rail switching	Manual Handling	
Element	A	D	E	H	A	
Distribution cycle 8	Rail only, unitized or palletized					
Test type	Handling	Vehicle stacking	Stacked vibration	Rail switching	Handling	Warehouse stacking
Element	B	D	E	H	B	C
Distribution cycle 9	Rail and motor freight, not unitized or palletized					
Test type	Handling	Vehicle stacking	Vehicle vibration	Rail switching	Loose-load vibration	Handling
Element	A or B	D	G	H	F	A or B
Distribution cycle 10	Rail and motor freight, unitized or palletized					
Test type	Handling	Vehicle stacking	Stacked vibration	Rail switching	Handling	Warehouse stacking
Element	B	D	E	H	B	C
Distribution cycle 11	Rail, TOFC and COFC					
Test type	Handling	Vehicle stacking	Rail switching	Stacked vibration	Loose-load vibration	Handling
Element	A or B	D	H	E	F	A or B
Distribution cycle 12	Air (intercity) and motor freight (local), over 100 lb, unitized or palletized					
Test type	Handling	Vehicle stacking	Stacked vibration	Vehicle vibration	Handling	
Element	A or B	D	E	G	A or B	
Distribution cycle 13	Air (intercity) and motor freight (local), single package up to 100 lb (45.4 kg)					
Test type	Manual handling	Vehicle stacking	Loose-load vibration	Vehicle vibration	Manual Handling	
Element	A	D	F	G	A	
Distribution cycle 14	Warehousing (partial cycle to be added to other cycles as needed)					
Test type	Handling	Warehouse stacking				
Element	A or B	C				

Figure 31 Distribution cycles for the selection of laboratory tests. (American Society of Laboratory Tests. American Society for Testing Materials, Committee D-10, used with permission.)

selected for sequential tests that will evaluate pack and unitizing adequacy for any product, in simulation of its distribution pattern (Figure 31). In formulating the battery of tests for each distribution pattern, there are defined to be eight elements in distribution which can develop stress hazards on packaged and unitized goods: manual and mechanical handling; warehouse and vehicle stacking; loose-load and unitized-load transportation; vehicle vibration and rail switching.

To simulate the stresses which can develop, programmable shock and vibration testers have been made available commercially and used to develop the protocols by which the battery of tests for any distribution cycle is executed. About 20 laboratories in the United States are equipped to comprehensively perform the battery of tests on shipping units up to unit loads in size. There is no doubt that the growth of this testing technology is reducing both damage losses and unnecessary overpacking for those packers who use it.

With results in hand on lab tests made with one or two unit loads—the tests need not be carried to destruction—the packaging engineer can specify means of stabilizing the unit loads against vibration at their natural frequencies, which has been recognized as the cause of most "hidden damage."

A unit-load spec can be worded, for instance:

Pallet pattern:	023
Tiers:	4
Natural Freq:	16.5 (cycles per second)
Stabilization:	Vertical plastic strapping

In some cases, a least-cost stabilization might consist of stretch-film wrap on the upper tier or tiers of a unit load, or even over-the-top tape if the primary packages fit without headspace into the shippers. In any case, low-frequency vibration (roughly in the range of 10 to 25 cycles per second) is a major hazard to unitized shipments, and the packaging engineer must be aware of its effect on the goods for which he specifies the packages and how to stabilize unit loads against it.

10

Computerized Specifications

All of the preceding material is "traditional" in format, i.e., the end products, namely specifications, are comfortably typed, drawings inserted or appended, reproduced on copiers, and distributed by mail. Originals are kept on file, from which more copies can be made at will for any valid reason, such as to solicit bids from a potential new supplier, or to start up a new plant.

A tidal wave of computer applications (which in lay terminology includes word processing) to all industrial functions requires that consideration be given to their role in the field of packaging specifications. As of this writing, a number of companies, probably not a large number, have applied computers—a very broad term—to packaging, which could mean materials inventory control, CAM (computer-assisted manufacturing), CAD (computer-assisted design), sales-purchasing coordination, cost analysis, and distribution control.

In keeping with the intent, as stated in the Preface, of focusing on packaging specifications as a communications medium, and quality-control base, it will be helpful to eliminate study of the research and development phase. Thus, we start at the point where the spec for a given package is defined, to identify the relative merits of "traditional" versus "computerized" specs.

When one for the first time takes the position that there is an option to computerize, or is told that "the company is computerizing" and packaging must fit in, a list of questions surfaces almost automatically. This becomes a time for cool heads and clear thinking, when an ounce of philosophy and a lot of precise definition can save years of frustration and misdirected effort.

Question	Translation
•What is to be accomplished;	•Who expects what results;
•Why do it?	•What justification, benefits, at what costs?
•How to do it?	•What hardware, software, and manpower are required?
•For whom?	•What business functions need, want, or would be helped?
•By whom?	•What business function controls and implements?
•How soon?	•How to avoid disruption?

The questions can be summed up as "What and when do you mean when you say 'computerize'?" It immediately becomes clear that there are many options, rather than one, and an analysis starting with the simplest shows that it's not simple.

I. OPTION 1. WORD PROCESSING ONLY

It is popular to replace typewriters with word processors; the document appears on a screen as it is being typed and is simultaneously recorded (or a short while later) on a disk, which becomes the original file copy. At any time, the disk can be inserted into a printer, which types a "hard copy" on paper at high speed; the result looks no different from a typed original. This describes a "stand-alone" system: it consists of a terminal (keyboard), CRT unit (screen), and printer (with disk drive).

A disk, though small, acts as a file for many documents; it is searched or scanned to find a desired document by mounting it on the disk drive in the printer and running it by on the screen, controlling the movements by keys on the terminal.

COMPUTERIZED SPECIFICATIONS 205

Looked at this way, the stand-alone word processor is a refined typewriter with features of a TV and record player: the document appears on the TV screen; and the disk is run on the equivalent of a turntable for both recording and playback. A person experienced in handling an electric typewriter can learn to operate a stand-alone word-processor system in 3 working days.

If this is what is meant by "computerizing" specs, it implies that present typists need 3 days' training on the new hardware, after which they reproduce on an office copier as before from the first printed "hard copy" and distribute by mail—with one exception; a word processor cannot handle prints, which are part of most specs. A file of prints must be separately maintained and appended for distribution, as in the traditional system.

Since the printer runs much faster than a typist, one printer can be switched among two or more terminals/CRT units; thus, a typist can be typing a document while another operator is printing from another disk, or scanning a disk to locate a document. If in a given company, packaging engineers are accustomed to looking in files for documents, they as well as secretaries and typists will have to be trained to locate documents from disks.

By way of appraisal, it can be said that although the stand-alone word processor is often one's first introduction to computer hardware, it is not a computer and can rarely be justified as a permanent installation. It requires higher skills than a typewriter, and costs more originally and for maintenance; its advantages are more-compact filing and somewhat greater security of information, if hard-copy files are in fact eliminated. When properly managed, a spec can be altered (and recoded for unique identification) by recalling the information stored on the file disk and inserting the new information; thus, only the current spec is on file.

The packaging function, in all but the smallest businesses, is always performed by a network of people, usually in diverse locations ranging from separate offices and departments in one building to nationwide or worldwide dispersal. Communication of spec issuances via word processors in these environments is worth some consideration. Starting with the simplest case, one can postulate a single-location business of moderate size, wherein all executive functions are centralized in a large building, and all manufacturing is done in adjacent buildings, including warehousing and distribution.

Assuming a reasonably sophisticated organization, the departments/functions which will need copies of the packaging specs include:

- Purchasing
- Production Control
- Engineering
- Research and Development
- Cost
- Distribution
- Packaging

Today's environment relative to installation of word processors and computers is a very fluid one and will probably remain so in the forseeable future: a great mix of hardware choices in many stages of complexity and compatibility among one another, with weekly introduction of new items in both hardware and software. Thus, our XYZ Company, as described above, may be in any of several situations, ranging from:

- Each department independently responsible for developing its own system, as it perceives its own needs, to
- All departments in step, according to a centrally controlled plan for development of word and data processing.

It is possible, in the first extreme, that all departments may have different and incompatible hardware. In that situation, the packaging specs may just as well be continued in the traditional pattern of typing, reproducing, and distribution by hand or interoffice mail, or prepared by the department's stand-alone word processors in the same manner.

Most likely, however, in a single-location company, central authority would be strong enough to require one source for the hardware in all departments, resulting in a high degree of compatibility. In this situation, the units in all departments can be linked to one another by direct wiring, such that a packaging spec can be called up on the CRT screen in any department and printed off in any location in the network. A couple of conditions are necessary: the appropriate disk (or diskette) holding the file copy of the desired spec must be available on a disk drive, and if a hard copy is to be made by the receiving department, a printer there has to be free.

The two conditions are more easily stated than executed. Mounting the desired disk can be arranged by an interoffice phone call, subject only to the normal hazards of completing interoffice calls, plus the condition that the call receiver is adequately trained to respond and has the time to do so immediately. Problems in achieving satisfaction via this route have led to two alternatives: accumulate requests for processing at a time of each day mutually agreeable to all departments (usually at the beginning or end of a normal office workday), or centralize disk files in one location for the whole company, with a couple of personnel to handle all requests. Given the adequacy of their training and their full-time concentration, the latter alternative requires a company-wide filing system to make all documents identifiable and retrievable.

This is a good point to consider another important discipline, namely, the authority to modify (and issue) packaging specs, as distinct from simply calling-up and making copies. In context, it is normal practice that each of several departments has the unique authority over certain kinds of documents; e.g., Research and Development for product specs, Purchasing for procurement contracts, Production Control for manufacturing schedules, Packaging for packaging specs, etc. The network must contain safeguards to allow only the word-processing unit in Packaging to modify a spec on file, while all other departments can call up and copy only. And so on reciprocally, with respect to the other departments in the network: Production Control must have the sole authority to alter manufacturing schedules, for another example, while Packaging may want a look at a given schedule to fit in a test.

All the problems implicit in the above discussion are capable of solution and are best solved by prevention, that is, by good planning before installation of hardware. That planning must include a full definition of the results required of the intended system. Two principles are paramount in considering the word-processing system for packaging specs:

- The Packaging Department must develop the plan in concert with all other departments concerned, to define the intended results.

- The established system, to be of value, should help packaging specialists to accomplish their work, without intruding on their time for the mechanics of word processing, beyond training for acclimation.

Another feature of word processing to be considered is that it is expandable, both in hardware and in software. Hardware expansion consists of adding units to a network—more stations at different locations, and terminals that will perform different functions. Software expansion most commonly means preprogramming: an example pertinent to packaging specs is the "glossary." In the past, it was called a "form"—a page or more with headings and/or a column or so of titles, leaving a series of blanks to be filled in by the specialist. Using the traditional method, a packaging specialist would pull from a desk drawer a packaging-spec form, which states all the criteria to be filled in; he penciled in the specifics, and a secretary or typist took over. Now, with a word processor, the packaging specialist can call up a glossary on the CRT screen, fill in the specifics, and file the resulting document on a diskette, for office personnel to handle from there.

System expansion to remote locations (beyond the direct-wiring range of 1000 feet or so for a network of word processors) raises complexity by involving telephone connections as part of the network. The term "telephone connections" includes microwave and satellite communications when distances exceed a couple of hundred miles between units in a network, as in a business with headquarters and plants, offices, and warehouses separated across a resgion or the nation.

In such an expanded environment, the word-processing network is a high-technology version of the traditional telex system, wherein a communication typed in one location is printed out in a remote location. If one wishes to transmit a packaging spec from headquarters to a distant plant, a terminal with keyboard and phone handset fitment is required at each end. The appropriate disk is mounted on the disk drive at the sending end; a CRT unit and/or printer is readied at the receiving end. The sender dials the proper phone number, and when the receiver picks up, both connect their handsets to the equipment, and the sender starts the disk drive.

The procedure just described illustrates the principle of transmission, but is in fact little used, because it is considerably more expensive than making a copy and mailing it from the headquarters. The volume of business communications is such that long-distance phone transmission of documents at speeds limited by printers would be a prohibitive expense. The common solution is to accumulate each day the documents to be transmitted and send them overnight, when rates are reduced, and in compressed form, that is, between terminals which electronically speed up the transmission up to 20 times normal, so that material which would normally take a minute to transmit can be sent in 3 seconds. This, of course, requires additional equipment at each end, which can be justified in a centralized corporate communications department that handles all the company's "traffic."

In summary, it is possible and feasible to use word processors for writing, filing, and distributing packaging specifications, and to modify them with less work than the traditional system of typing and copying. It is difficult to justify such a system economically for a small business or a single-location business, unless the business as a whole is adopting word processing company-wide for all functions. In general, the larger the company, the larger the intracompany traffic in communications, and the easier to justify. All the normal precautions for control of specifications, authority for issuance and change, and distribution must be exercised in word processing as in the traditional mode. Planning the system and network before starting to implement is critical, to be sure both that the needs of all functions and departments concerned with packaging are included, and that hardware and software compatible company-wide are selected. Finally, operation of a word-processing system should be designed to increase the effective use of time by packaging professionals on their priority objectives.

II. OPTION 2. DATA PROCESSING

A complete set of packaging specifications for a business represents a great deal of information, but much of it is indirect. For example, it represents the total bill paid for packaging material, but only when multiplied by the production volume for each spec and the price per 1000 packages. It represents the tonnage subject to outbound freight, when multiplied by the production volume and the gross weight per packed shipping unit. The specs can be sorted by code to indicate dependency by weight or cost on any given packaging material or ingredient, when multiplied by production volume. The difference between volume of a given spec purchased and volume shipped reveals internal shrinkage or overorder—both valuable management information. If an ingredient in a packaging spec is raised in price by 10%, say, the total incremental cost for a future time period can be determined by identifying the affected specs and planned production volumes.

Manipulating the information in packaging specs to obtain useful and usable conclusions has traditionally been done "by hand," that is, by selecting a few specs known by the Purchasing Department or Production Control to be used in large volume and estimating effects of any change on the whole based on examining a "major part."

To cover all specs, or even to identify those which are affected in a company which uses thousands of specs, is too time-consuming to be practical "by hand." But such manipulation is entirely practical with data-processing equipment, *if* the specs are set up to include the information to be manipulated.

It is therefore imperative to build into the spec format and content all of the items that all departments concerned with packaging will want fulfilled in data processing. Examples of some items are:

Purchasing, to estimate effects of vendor activities
- Suppliers of a given spec, by plant
- Specs using a given component, or ingredient

Production Control, to plant anticipated needs
- Summation of production volumes by spec and plant
- Comparison of planned volumes versus plant capacity

Engineering, to plan facilities for production
- Packaging machinery required by spec
- Line speeds by spec

Research and Development, to relate product criteria
- Product shelf life, spec targets by product
- Product compatibility
- Regulatory compliance

COMPUTERIZED SPECIFICATIONS 211

Cost, to base financial needs
  • Estimate inventory costs for packaging materials
  • Set up base for measuring production yields and shrinkage
Distribution, for inventory and freight management
  • Gross weight, by spec, finished goods
  • Damage criteria in re: claim handling

In prior chapters it is illustrated how specifications show structure and performance in a traditional system; given the discussion to this point on "computerized" specs, would the format and content differ, and if so, how? The requirement for adaptability to data processing alone would indicate the necessity of some differences.

A listing of the criteria to be met helps to define the computerized spec:

1. Unique identity is an absolute requirement; that is, the spec coding system must not be ambiguous. This is no different from a traditional spec format.

2. The coding system must be built for long-term use, that is, capable of including added items almost indefinitely, like a telephone system, within a basic logic framework.

3. Data processing will require many different "cuts" when specs are scanned by different departments for different purposes; thus, they had best be modularized so that the separate package components can be readily dealt with. This was done to some extent in the traditional specs of prior chapters, but may benefit from finer cuts in the computerized spec. For instance, the can spec (Chapter 6) is numbered 520810 and includes both body and ends. A can end is really a separate package component, and its structure may be altered without changing the body, or vice versa. In a computerized spec, where some of the information is subject to data processing, it would be advisable to assign a separate code to the end spec, for easier access to information on the effect of potential change in either end or body.

4. As in traditional specs, it is advisable to avoid inclusion of graphic specs (design numbers), since several designs are commonly applied on one structure, e.g., several flavors in a line of products, each in the same carton structure. Graphic defects which are generic (not specific to a given design, such as "missing color," "off-register," "scratched," "inadequate scuff resistance") should be included where appropriate, as in specs for cartons, lithographed cans, and printed labels.

5. Spec format as well as coding system should be uniform company-wide. This would seem too obvious to require statement, but decentralized corporations have been known to exercise *laissez faire* to the extent that specs for identical packaging issued by different divisions could not be recognized as such. Varying formats make data-processing efforts more difficult at best and can prevent centralized purchasing functions from combining volumes for price breaks in contracting.

6. A follow-on to the preceding—specs should state dimensions and tolerances *standardized* for all divisions and plants, using industry standards whenever possible. For instance, can-making metals have industry-wide tolerances for weight and temper on every commercial spec, injection-moulded plastic components have commercial tolerances for dimensional variation, paperboards have commercial tolerances for gauge and density. If different divisions or plants of the XYZ Corporation insist on deliveries with half the commercial tolerances, that is in fact a separate spec from the same made to commercial tolerances and carries a premium price. Volumes for all plants cannot be combined or cumulated for purchasing, and supplies are not interchangeable among plants; this is the kind of situation that leads to the much publicized examples of the military procurement system, which pays many times the commercial price for tight-tolerance specs.

7. References to quality-inspection test methods should be included. They may be ASTM, TAPPI, GMI, etc. methods where applicable, or coded references to methods agreed upon between vendor and buyer.

8. Ideally, each spec should be presentable on the equivalent of one ordinary page. Thus, a complete package is set forth in a group of specs: one page for the retail container or for each component thereof, one page for the shipping container, and one page for the unit load, plus one page for containerization, if applicable.

9. Prints are over and above the aforesaid, usually referred to by number and separately distributed.

10. Given the systematized format thus far described, the amount of text in specs can be minimized in favor of tabulated information.

COMPUTERIZED SPECIFICATIONS 213

At this point it is appropriate to see how a few computerized specs would look. The examples shown are based on industrial practice developed over the period 1978 to 1984. The organization of a complete package spec consists of about one page for each component, the whole tied together with a first-page summary.

The first exhibit (Figure 32) shows the glossary or form for the Package Summary. This is a software item, which the package developer calls up on the screen of his work station (or the department secretary does it) when a spec is ready to issue. The second exhibit (Figure 33) shows the same, with blanks filled in to describe a 12-ounce liquid detergent package group of specs. The reader will note that the Summary is simply a key to all component specs. Separating the components thus makes it easier to cover with each component buyer (and supplier) only the information pertinent to their specific interests and activities.

Referring to the material in Chapter 1 on *numbering* or, more properly, *uniquely identifying* specs, it would be worthwhile to explain the alphanumeric system illustrated by the spec and drawing numbers in Figure 33. "Alphanumeric" means a mixture of letters and numbers, which is today easily handled in both word and data processing. The combination must, of course, be determined in accordance with a system. The specific can be arbitrarily decided within the user company; the sole requirement is that the system be consistent across the company, but need not be consistent with any other company (and should not, for security reasons).

An analogy on a larger scale is mail codes: The U.S. Postal Service uses a five-digit number system which is consistent for the 50 states, while Great Britain uses a six-bit system in which the third and fourth are numerals and the rest are letters. Canada also uses a six-bit system, but there the odd positions are letters and the even are numerals.

Thus, the data in Figure 33 illustrate *a* system, but by no means *the* system for spec identification. The first step is to assign meaning to each of the positions in the total code: in this case

```
XYZ COMPANY                                    SPEC. NO.
PACKAGING MATERIAL SPECIFICATION               DIVISION
PACKAGE SUMMARY                                EFF. DATE
PRODUCT
PACKING PLANTS
DESCRIPTION
PRIMARY PACKAGE COMPONENTS
  DESCRIPTION                       SPEC. NO.         DRAWING NO.




SHIPPING CONTAINER COMPONENTS
  DESCRIPTION                       SPEC. NO.         DRAWING NO.




UNIT LOAD/DISTRIBUTION INFORMATION
  DESCRIPTION                       SPEC. NO.         DRAWING NO.
```

Figure 32   Sample specification: Glossary for package summary.

COMPUTERIZED SPECIFICATIONS

```
XYZ COMPANY                                   SPEC. NO. PMDDL12945650
PACKAGING MATERIAL SPECIFICATION              DIVISION  Detergent
PACKAGE SUMMARY                               EFF. DATE 01/07/85
PRODUCT            Pristine Brand Liquid Detergent, cases of 24
PACKING PLANTS     1,4,5,7,10
DESCRIPTION        12-oz plastic squeeze bottle, paper lables, disp. cap
PRIMARY PACKAGE COMPONENTS
DESCRIPTION                          SPEC. NO.         DRAWING NO.
Component A
 Extrusion-blown bottle, poly-
 propylene copolymer, 12 oz., 28-400 fin.  945651       BPC51870

Component B
 Injection-moulded closure, push-pull,
 HDPE, 28-400 thread                  945652           CPC20664

Component C
 Paper labels, front and back, water-
 resistant overprint lacquer          945653           LC053551

Component D
 Label glue, PVA dispersion           AWP005           none



SHIPPING CONTAINER COMPONENTS
DESCRIPTION                          SPEC. NO.         DRAWING NO.
Component A
 RSC corrugated, for 24 bottles,
 200T, C flute                        945654           CCC75463

Component B
 Partitions, corrugated, 200T, C flute,
 3 long and 5 short                   945655              "



UNIT LOAD/DISTRIBUTION INFORMATION
DESCRIPTION                          SPEC. NO.         DRAWING NO.
Load pattern/platform                IL010/slip sheet  same

Load stabilizer                      none needed       n.a.

Clampable                            no                n.a.

Tiers per unit load                  4                 n.a.

Tiers per stack, max.                16                n.a.

Case cube/weight/cases per unit load 1.2cf/27 lb/36    n.a.

Slip sheet size/weight               48 x 40/2.0 lb    L2ASK
```

Figure 33 Example of a filled-out package summary.

Positions	Communicate
1 and 2	Kind of spec (PM = Package Material)
3	Division (D = Detergent Div.)
4 and 5	Product type (DL = Detergent, Liq.)
6 and 7	Product no. (12 = Comm. formula)
8 thru 13	Package no. (mixed classification, per Chapter 1)

The first two positions distinguish a package spec from a product spec, a process spec, a machine spec, test method spec, etc., which could be designated by PD, PS, MA, TM, etc., respectively, to cover as many kinds of semipermanent documents as the company issues, including policies, procedures, GMPs, etc.

The Spec Nos. 945651 through 945655 are really subspecs of 945650, tied to Product L12 of the Detergent Division. Thus, Components A and B of the primary package could call for pigment coloration in their respective detailed specs, because their full designations are PMDDL12945651 and 2. Should the Detergent Division use the same package, but different colors, for another product (brand) named "Household," and coded L13, the specs for the bottle and closure would be PMDDL13945651 and 2, respectively.

The number of bits for each category depends on the size of the category. If a company has 26 or fewer kinds of documents to reference, the first category could be covered by a single letter, rather than two. If there are more than 26 divisions, and less than 100, two numerals or two letters would suffice, and so on.

As for the Drawing Nos., the first three letters indicate component form and material, e.g.;

Letters	Communicate
BPC	Bottle, plastic, polyolefin
CPC	Closure, plastic, polyolefin
LCO	Label, cellulosic, overlacquered
AWP	Adhesive, waterproof
CCC	Corrugated case, C flute

The numbers following the letters are sequential.

COMPUTERIZED SPECIFICATIONS

```
XYZ COMPANY                                      SPEC. NO. PMHPC33511071
PACKAGING MATERIAL SPECIFICATION                 DIVISION  Household Prod.
PRIMARY PACKAGE                                  EFF. DATE 02/15/85
PRODUCT        DecoReady Patching Compound Mix, 125 grams
PACKING PLANTS 1 and 5
DESCRIPTION    Component A, Laminated Pouch
```

STRUCTURE (outer to inner)	WEIGHT, Lb/3M ft^2 Nominal	± Tolerance	GAUGE, Mils Nominal	± Tolerance
Scuff/gloss coating	2.5	0.1	—	—
Supercal. sulfite pouch paper	26.0	1.3	1.7	0.1
LDPE laminant, 8 m.i.	15.0	1.5	1.0	0.1
Aluminum foil	12.6	1.3	0.3	0.03
LDPE seal, 8 m.i.	24.0	2.4	1.5	0.2
Total	80.1	6.6	4.5	0.43

FORMED PACKAGE: Horizontal f/f/s, IM or CM equipment, 4-side seals

Dimensions	Nominal inches	± Tolerance inches	Test Method	Defect Class
Height	7.5	0.125	TM401	B
Width	4.9	0.098		
Seal Widths				
Sides	0.25	0.063		
Bottom	0.13	0.032	TM401	B
Top	0.50	0.125		
Incomplete seals			TM450	A
Ink pickoff, heat seal areas			Visual	C
Seal strength, 350°F, 0.5 sec., 40 psi: destruction			TM225	B

ROLL STOCK PROPERTIES

WVTR flat/creased	0.05 gm/0.05 gm maximum	TM106	A
Stiffness, Taber units	3.0 MD/2.2 CD minimum	TM122	B
Laminate bond strength	200 gm/inch minimum	TM104	B
Scuff resistance	#2 rating, maximum	TM110	C
Web width	15.00" ± 0.0625	TM101	B
Eye spot repeat, 1 impr.	4.88" ± 0.0313	"	B
" " " 10 impr.	48.75" ± 0.0625	"	B
Area per impression	73.125 sq.in. nominal	TM102	—
Impressions/lb.	72.3	TM103	B
Max. roll diameter	15.0 in., incl. 3" core	—	A
Impressions per roll	6860	—	B
Max. splices/roll	5	—	B
Roll weight	95 lb.		

DELIVERY

Unit load	48 x 40 Grocery pallet, 4-way entry, wood
Roll count/pattern	24 rolls, 8 rolls per tier, 3 tiers column stacked:
Load stabilization	plastic straps $\frac{\overline{X}\,\overline{X}\,\overline{X}}{X\,X\,X}$
Sanitation	each roll sealed in PE film

Figure 34 Sample specification for a heat-sealed pouch.

In the Unit Load Information section, there is too little detail to justify a separate detail spec except for the pattern in which the unit load (also called "module") is assembled. The term "IL010" means "Interlocking Pattern #10" in a library of a couple of hundred patterns by which a group of cases may be compactly laid out on an area about 48 inches × 40 inches. One of the earliest packaging applications for computer software was the program for selection of unit load patterns, starting with horizontal dimensions of any given case. The software has been refined to capability for resolving inefficient patterns to achieve the maximum of area utilization.

The drawing referred to for the slip sheet is a generic "Lip 2 Adjacent Sides—Kraft" structure, as available from a number of sources with minor differences. The item is disposable, and purchasing flexibility would normally take priority over detailed specs and drawing.

Next, to look at a detailed spec in tabular form, Figure 34 describes a four-side-seal pouch of paper/LDPE/aluminum foil/LDPE. This is a primary package based on a workhorse structure of over 30 years' volume production, with the principal evolution of foil gauge from 0.00035 inch to 0.00030 inch. The third position in the identifying code is "H," representing the Household Products Division; the fourth and fifth are "PC" for Patching Compound; the sixth and seventh are the product formula number; and the rest are the package number. Overall, this spec does on one page what the conventional spec in Chapter 8, Polyethylene Bread Bag, No. 72160, takes three pages to say (exclusive of drawing). Part of this compression results from referring to test methods, rather than stating them; otherwise, it doesn't require computerization to reduce a spec to one page. Computerization favors abbreviation for information storage and retrieval, and if abbreviation is good for the conventional system, it can be applied there also.

Appendix:

Technical and Trade Associations

AIMCAL	Association of Industrial Metallizers, Coaters and Laminators 61 Blue Ridge Road, Wilton, CT 06897
ASTM	American Society for Testing Materials 1916 Race St., Philadelphia, PA 19103
FBA	Fiber Box Association 5725 East River Road, Chicago, IL 60631
FPA	Flexible Packaging Association 1090 Vermont Ave., N.W. Washington, D.C., 20005
GPI	Glass Packaging Institute 6845 Elm St., McLean, VA 22101
NPBA	National Paper Box Association 231 Kings Highway East, Haddonfield, NJ 08033
PI	Packaging Institute, U.S.A. 20 Summer St., Stamford, CT 06901

PPC   Paperboard Packaging Council
      1101 Vermont Ave. NW, Washington, DC 20005

SPHE Society of Packaging and Handling Engineers
      Reston International Center, Reston, Virginia 22091

TAPPI Technical Assn. of the Pulp and Paper Industry
      One Dunwoody Park, Atlanta, GA 30341

Bibliography

CHAPTERS 1 AND 2

1. 7 Steps to successful packaging, *Modern Packaging Encyclopedia*, Vol. 43, No. 7A, pp. 36-38, (July 1970).
2. *A Basic Guide to Preparing Packaging Specifications*, Special Report No. 8, American Management Association, New York (1956).
3. How to cope with military specs, *Mod. Packaging*, pp. 103-105 (Sept. 1969).
4. R. Hottinger and K. Bence, Specifications, tests cut shipping damage, *Package Engineering*, pp. 103-108 (June 1965).
5. C. Koehler, Maintaining proper packaging specifications, in *New Techniques for the Packaging Engineer*, Packaging Institute Technical Publication (1954).
6. K. F. Lang, Over $1 million reduction in packaging costs at Heinz, *Food Processing*, pp. 36-37 (June 1957).
7. I. R. Linnard, "What Top Management Expects from Specifications and Inspection of Packaging Materials," Address, Sept. 22, 1960 (Lambert-Hudnut Mfg. Labs, Inc.).
8. G. G. Maltenfort, Performance or material specifications? You need both, *Package Engineering*, pp. 64-66 (June 1968).

9. T. Neill, How to write and keep an up-to-date spec sheet, *Can. Packaging*, pp. 78-79 (April, 1970).
10. W. Stern, *The Package Engineering Handbook*, Board Products Publishing Co., pp. 217-227 (1949, 1954).
11. E. A. Leonard, *How to Improve Packaging Costs*, AMACOM Div., American Management Association, New York, (1981).
12. W. K. Fallon, *AMA Management Handbook*, 2nd ed., AMACOM Div., American Management Association, New York, Chapter 13 (1983).

CHAPTERS 3 AND 4

1. C. H. Bubb, *Quality and Its Control*, Bulletin 106, American Management Association, New York (1967).
2. A. F. Cowan, *Quality Control for the Manager*, 2nd ed., Pergamon Press, New York (1966).
3. A. F. Deuble, Quality control enters the packaging department, in *New Techniques for the Packaging Engineer*, Packaging Institute Technical Publication (1954).
4. H. A. Freeman, *Industrial Statistics*, John Wiley, New York (1942).
5. J. M. Juran, ed., *Quality Control Handbook*, McGraw-Hill, New York (1962).
6. E. A. Leonard, *Economics of Packaging*, pp. 1-13: Harcourt Brace Jovanovich (1980).
7. W. Stern, *The Package Engineering Handbook*, Board Products Publishing Co., pp. 233-242 (1954).
8. U. S. Dept. of Defense, *MIL-STD-105D, Sampling Procedures and Tables for Inspection by Attributes*, Supt. of Documents, U.S. Government Printing Office, Washington, D.C. (April 29, 1963).
9. *Checking Prepackaged Commodities, National Bureau of Standards*, Handbook 67, U. S. Dept. of Commerce, U.S. Gov't. Printing Office, Washington, D.C.
10. R. E. Seely, A procedure for estimating the quality of a production lot, *Package Dev.*, p. 14 (Nov./Dec. 1975).
11. E. F. Daigler, Establishing a statistical weight control program for on-line use, *Package Dev.*, p. 19 (May/June 1976).

CHAPTER 5

1. B. Deitch and D. Collett, Glass containers and metal STC closures, *Aust. Packaging*, pp. 69-71 (July 1968).
2. R. D. Dubble, *Why Surface Coatings?* Anchor Hocking Glass Co. (July 1, 1965).

BIBLIOGRAPHY

3. B. E. Moody, *Packaging in Glass*, Hutchinson, London (1963).
4. *Glass Containers 1970*, Glass Packaging Institute, McLean, Virginia (1970).
5. *Standard Glass Container Series 1*, Glass Packaging Institute, McLean, Virginia (June 1956).
6. *Glass Defects*, Hartford Empire Company (1957).
7. *List of Aids to Inspection*, Special Report 61-2, Packaging Institute, U.S.A., Stamford, Connecticut (June 1961).
8. *Closure Sizes and Gages*, Recommended Specification 0-2r-61, Packaging Institute, U.S.A., Stamford, Connecticut (June 1961).
9. T. W. Dickes, Glass-to-glass packaging, *Package Dev.*, p. 12 (May/June 1975).
10. C. D. Gray, Corrugated and Its Interface with Bulk and Partitionless Packaging, Proceedings of the 1977 International Packaging Week Assembly, Paper No. 7710 (The Packaging Institute, U.S.A., New York, N.Y.).
11. *Secondary Glass Packaging—Voluntary Specifications Guideline*, National Soft Drink Association, Washington, D.C. (1979).
12. Timesavers—Glass finishes, *Package Dev., p. 28 (Jan./Feb. 1976)*.

CHAPTER 6

1. R. R. Hartwell, "Choice of Containers for Various Products," Address, Third Int'l. Congress on Canned Foods, Rome, American Can Co. (Sept. 1956).
2. *Tin Mill Products*, United States Steel Corp. (Sept. 1968).
3. H. S. Cannon, "The Tin Free Steel Revolution," Proceedings of the 30th Annual National Forum, The Packaging Institute, U.S.A. (Oct. 1968).
4. A. Serchuk, A new coat for cans, *Mod. Packaging*, p. 36 (Sept. 1979).
5. Trends in steel can coating/curing: Staff report, *Food Processing*, p. 86 (March 1982).
6. C. Andres, Food processors benefit from 2-piece vs. 3-piece can technology, *Food Processing*, p. 124 (June 1981).
7. C. Andres, New bottom bead configuration improves 2-piece can, *Food Processing*, p. 58 (March 1980).
8. Tin-free steel cans move forward: Staff report, *Food Engineering*, p. 136 (Dec. 1981).

CHAPTER 7

1. Introduction to plastics, *Modern Packaging Encyclopedia*, Vol. 43, No. 7A, pp. 121-122 (July 1970).
2. Characteristics of plastics for rigid and semi-rigid containers, *Modern Packaging Encyclopedia*, Vol. 43, No. 7A, pp. 126-128 (July 1970).

3. R. G. Taylor, The plastic bottle, *Modern Packaging Encyclopedia*, Vol. 43, No. 7A, pp. 266-269 (July 1970).
4. G. J. Kundrat, Plastic tubes in packaging, *Modern Packaging Encyclopedia*, Vol. 43, No. 7A, pp. 270-271 (July 1970).
5. T. E. Neely, Plastic closures, *Modern Packaging Encyclopedia*, Vol. 43, No. 7A, pp. 288-294 (July 19700.
6. B. Carow, Dispensing closures, *Modern Packaging Encyclopedia*, Vol. 43, No. 7A, pp. 295-297 (July 1970).
7. J. H. Briston, *Moulded Thermoplastics Containers: Packaging Materials and Containers*, Blackie and Son, Ltd., London, pp. 194-210 (1967).
8. G. Bell, Standard tests for plastics, *Plastics World*, p. 60 (April 1980).
9. S. Medlock, Blow molding innovations, *Packaging Engineering*, p. 31 (Dec. 19820.
10. Staff report: Coextrusions hurdle barrier problems, *Packaging Digest*, p. 86 (Nov. 1982).

CHAPTER 8

1. *Modern Packaging Encyclopedia*, Sections 3, 4, 5, 6, and 9, Vol. 43, No. 7A (July 1970).
2. R. L. Harding, Jr., *Materials Guide—Selecting Films for Packaging*, Fact Sheet Supplement #2, The Packaging Institute, New York (1968).
3. *State of the Art: Processing Plastics for Packaging*, Plastics Technology: 17, 1, Bill Bros. Publ., New York (Jan. 1971).
4. *The Folding Carton*, Folding Paper Box Association of America, New York, Chicago, Los Angeles, (1950, 1954).
5. C. E. Price, Control of folding carton scorebend, opening force cuts spoilage, hikes efficiency of packaging lines, *Package Engineering*, 10: 100-109 (May 1965).
6. Staff article, Package engineering at general motors—Setting the specifications, and controlling the output, *Package Engineering*, *12*:59-64, 77-82 (Jan. 1967).
7. *TAPPI Standards and Suggested Methods*, Technical Association of the Pulp and Paper Industry, New York (1966).
8. A. Lambell, *Specifications for Shipping Containers*, Australian Packaging: Bell. Publ., Sydney, pp. 50-53 (Sept. 1969).
9. G. G. Maltenfort, Specifications for corrugated shipping containers—Sense and nonsense, *TAPPI Journal*, *50*:70A-72A (June 1967).
10. P. Cope, Measuring and specifying bulge and crush resistance in cartons and carton board, *TAPPI Journal*, *44*: 633-633 (Sept. 1961).
11. W. E. Oates, Jr., A world of change—boxes too! *Prof. J. Packaging Inst.*, pp. 5-7 9Fall 1970).

BIBLIOGRAPHY

12. Staff report, Paperboard in food packaging, *Food Engineering*, p. 76 (Jan. 1977).
13. A. D. Griffin, Developments in dense paper specialties, *Package Dev.*, p. 21 (Jan./Feb. 1976).
14. R. Bannar, Metallized film, *Food Engineering*, p. 63 (Oct. 1977).
15. W. C. Spring, High barrier metallized films, *Paper, Film, Foil Converter*, (Feb. 1980).
16. J. H. Carter, Packaging experts turn to metallized papers, *Paper, Film, Foil Converter* (March 1982).
17. P. J. Clough, Barrier treated paper and board, *Packaging Technol.*, p. 20 (Aug. 1980).
18. Staff report: Flexible packaging, *Package Engineering*, p. 39 (March 1979).
19. J. Rice, Basic users' guide to packaging films, *Food Processing*, p. 65 (March 1982).
20. P. A. Absalom, Flexible packaging materials, *Packaging Technol.*, p. 17 (Dec. 1981).
21. R. A. Lampi et al., Performance of flexible package seals, *Mod. Packaging* (May/June 1976).
22. L. L. Scheiner, New approaches in liquid F-F-S, *Mod. Packaging*, p. 25 (Oct. 1978).
23. K. Bertrand, Bag-in-box striding forward materially, *Package Engineering*, p. 33 (June 1982).
24. W. H. Clifford and S. W. Gyeszley, Calculating pouch volumes, *Mod. Packaging*, p. 38 (Aug. 1977).
25. L. Brazier, Effects of adhesives on flex crack performance of flexible packaging, *Package Dev. Systems*, p. 22 (May/June 1978).
26. F. Sussenberger, Reduce folding carton costs by judicous structural design, *Package Dev.*, p. 25 (July/Aug. 1975).
27. L. J. Timmer, An effective tool for measuring and comparing folding carton quality, *Package Dev. Systems*, p. 27 (May/June 1979).
28. Staff report, New horizons on laser diemaking and carton design, *Paperboard Packaging* (Magazines for Industry, Inc.) (1981).

CHAPTER 9

1. M. Siegel, Shrink and stretch equipment, *Modern Packaging Encyclopecia and Packaging Guide*, Morgan-Grampian Publishing Co., pp. 144-150 (Dec. 1975).
2. Staff report, Casebook of pallet uses and developments, *Material Handling Engineering*, pp. 50-54 (June 1975).
3. Staff report, Giving pallets the slip, *Modern Packaging*, p. 27 (March 1975).

4. Staff report, Stretch wrap pallet loads..., *Food Processing*, pp. 48-50 (Sept. 1974).
5. Staff report,...The economics of pallet use, Easter Pallet Users Conference, *Mat'l. Handling Engineering*, pp. 67-69 (Feb. 1975).
6. Military Specification MIL-L-35078K/GEN: Unit Loads for nonperishable subsistence items: Dept. of the Army, Natick Research & Development Command, Natick, Massachusetts.
7. G. C. Richardson, Package testing to match the distribution environment, *Package Engineering*, p. 44 (Dec. 1980).
8. Staff report, Focus-package testing, *Packaging Technol.*, full issue (Dec. 1982/Jan. 1983).
9. ASTM D10-22, *Performance Testing and Shipping Containers*, American Society for Testing Materials, Philadelphia, (1981).
10. R. Promisel, "Slip Sheets vs. Pallets—the Decision-Making Process, Proceedings of the 1979 PI Western Regional Packaging Forum, Paper No. F-7914, The Packaging Institute U.S.A., Stamford, Connecticut.
11. C. W. Ebeling, Push-pull or pallets? Distribution's dilemma," *Handling Shipping Management*, p. 42 (Oct. 1978).
12. T. J. Urbanik, Effect of transportation vibration on unitized corrugated containers, *Package Dev. Systems*, p. 13 (Jan./Feb. 1980).
13. G. Tribble and R. Flaum, Picking the best strapping to unitize..., *Food Drug Packaging*, p. 31 (July 1983).
14. I. Dermansky, A quick guide to effective film stabilization of pallets, *Food Drug Packaging*, p. 23 (July 1983).

CHAPTER 10

1. Staff report, Computerization, *Package Engineering*, 20:40 (Sept. 1975).
2. The computer, *Package Engineering*, 40 (Jan. 1979).

Index

Alphanumeric identification, 9-10, 12, 14, 16, 213, 216

Bags, film, 153-154, 159, 161-164
  defects in, 162
  delivery of, 161-162
  inspection of, 163
  quality control of, 162-163
  specification of, 161-164
Blister packaging, 146-151
  materials, 146
  specification of, 147-150

Cartons, paperboard, 177-179
  defects, classification of, 178
  inspection of, 178
  specification of, 177-179
Closures:
  compression-moulded, 131

[Closures]
  injection-moulded, 63, 64, 65, 68-69
  metal, 89-92, 99
  inspection of, 68, 90, 106-107, 108, 137-138
  sampling, 68, 70-72
  specification of, 89-92, 137-140
Coextrusions, 164-167
  description of, 164-166
  inspection of, 166-173
Computerized specifications:
  data processing, 209-212
  glossary, 208, 213, 214
  identification, 213, 216 (*see also* Alphanumeric identification)
  word processing, 204-209
Corrugated shippers:
  quality aspects, 180-182
  specification of, 96-98, 120, 121, 145, 215

Criteria, 2, 6
   functional, 18-27
   for graphics, 26-27
   marketing, 6, 7

Defects:
   in blister packages, 148-149
   in cans, 114-115, 126
   in cartons, 178
   classification of, 38-39, 74-75
   in coextrusions, 173
   in corrugated shippers, 97, 120-121, 180-181
   in film bags, 162
   in glass, 85-86
   in labels, 94
   in metal closures, 90-91
   in overwrapping, 156-157
   in plastic bottles, 135
   in plastic closures, 138
Defectives, from plant operations, 33-41, 128, 130, 151
Delivery, from supplier to packer, 41-43, 51-52, 114, 123-124, 134, 137
Distribution packaging, 183-198

Flexible packages, 155-176
   film bags, 159, 161-163
   metallized packaging, 173-176
   overwraps, 155-160
   pouches, 217

Graphics, 3
   on cartons, 178
   on corrugated shippers, 97
   criteria for, 26-27
   on labels, 93, 94
   on overwrappers, 157
   silk screen, 133, 135
   in specifications, 211-212

Inspection:
   of blow-moulded containers, 134-135
   of cans, 63-64, 65, 68, 69-70
   of cartons, 178
   definition of, 60
   of glass, 104-106
   of injection-moulded parts, 137-138
   of pre-formed bags, 163
   in the quality control process, 57-59
   of roll stock (flexible materials), 157-158

Metallized packaging, 173-176

Overwrappers, 153-160

Pallets, 184-185
Pallet patterns, 186, 187, 189
Paper, waxed, 153-154
   defects in, 156-157
   inspection of, 157-158, 160
   specification of, 155-159

INDEX

Plastics packages, 131-151
  blow-moulded, 132, 133-136
  film packages, 153-154, 161-164
  injection-moulded, 132, 137-140
  thermoforms, 146-151
Purchasing, 3, 45-55
  buyer, 3, 50, 52
  contracts and orders, 46, 47
  in-plant shrinkage, allowance for, 41, 50-51
  usage of specifications by, 2, 3, 45, 49-51

Quality:
  attributes and variables, 8, 75
  of corrugated shippers, 180-182
  criteria, 43-44
  definition of, 73
Quality control:
  of blow-moulded containers, 139, 141-145
  of cans, 33, 37, 124-130
  of can ends, 126-128
  of glass, 33-34, 35, 37, 39-40, 101-106, 141-145
  of labels, 107, 109
  of metal closures, 106-107, 108
  of plastic closures, 139
  process of, 57-58

Sampling, 63-73
  of cans, 65, 68, 69-70, 115, 123
  of can ends, 127
  of closures, 64, 65, 68, 70-72
  of corrugated shippers, 121, 124
  of glass, 84
  of labels, 109
  of thermoforms, 151
Shipping containers, regulation of, 197
Shrinkage, from packing operations, 40-41
Slip sheets, 193, 195-196
Specification:
  of consumer function, 24-27
  definition of, 2, 3, 5-6
  formal, 4, 5, 6
  of product protection, 18-22
Specifications:
  of can ends, 112, 113, 114, 115, 118
  computerized, 203-218
  of corrugated shippers, 96-98, 120, 121, 124, 145
  development of, 4
  of glass, 81-88, 99
  identification of, 9-16, 213, 216
  of labels, 22, 25, 93-95, 99, 119, 122
  of metal cans, 111-118
  of metal closures, 89-92, 99
  of plastic closures, 137-140
  structural, 7-8, 29-31

Standards, 73-76, 79
  for glass, 99, 123
  for cans, 123

Test methods, 60-63, 127, 176, 198-201
Thermoforms, 146-151
  defects in, 148-149
  inspection, 148
  quality control of, 148, 151
  sampling of, 151
  specification of, 147-150

Unit loads:
  of cased plastic bottles, 215, 218
  definition of, 184
  of flexible roll stock, 217
  forms of, 191, 193-195
  performance of, 198-201
  stabilization of, 193-196, 201